ENTOMOLOGICAL SOCIETY OF AMERICA

Handbook of Corn Insects

EDITED BY

Kevin L. Steffey
Marlin E. Rice
John All
David A. Andow
Michael E. Gray
John W. Van Duyn

Entomological Society of America
9301 Annapolis Road, Suite 300
Lanham, MD 20706–3115 U.S.A.
(301) 731-4535
FAX (301) 731-4538

The Entomological Society of America (ESA) is a not-for-profit organization serving the scientific and professional needs of entomologists and individuals in related disciplines throughout the world. Formed in 1953 by the consolidation of the American Association of Economic Entomologists (founded in 1898) and the former Entomological Society of America (founded in 1906), ESA is the largest international association of entomologists. It has more than 7,900 members at colleges and universities, state and federal departments of agriculture, other government agencies, health agencies, private industry, parks, and museums.

Front cover: Southwestern Corn Borer photograph by Marlin E. Rice
Back cover: Differential Grasshopper photograph by Marlin E. Rice

The *Handbook of Corn Insects* was produced by
Alan Kahan (Director of Communications)
and Dawn Davern (Production Editor)
of the ESA Publications Department.

PRINTED AND BOUND IN THE UNITED STATES OF AMERICA

Contents

Contents

Acknowledgments

We have waited a long time to acknowledge all of the individuals who have given unselfishly of their time and expertise to help us complete the *Handbook of Corn Insects*. This handbook was "under construction" for a long time, for many different reasons, and we sincerely appreciate the patience of the authors and ESA membership as we aspired to produce a finished product of which we could be proud.

Like the editors of previous handbooks in the series, we thank former ESA President Lowell Nault, Ohio State University, for his vision for the handbook series and his efforts to initiate the series. We also thank the individuals who have served on the ESA Handbook Series Editorial Committee for their guidance and leadership. Their support during delays in preparation and production was much appreciated. Individuals from the committee who reviewed the entire manuscript and provided valuable comments and suggestions were David Boethel, Louisiana State University; Rick Brandenburg, North Carolina State University; Leon Higley, University of Nebraska; George Kennedy, North Carolina State University; Daniel Potter, University of Kentucky; and Victoria Yokoyama, USDA–ARS Horticultural Crops Research Laboratory, Fresno, CA. We appreciate their sacrificing their valuable time to provide much-needed input. We especially thank David Boethel and Leon Higley, the pioneer editors in this series of handbooks, for their empathy and encouragement.

Many individuals were involved in developing the final product. More than 70 authors contributed sections or chapters to this handbook. Their expertise, hard work, and willingness to accommodate the production schedule are sincerely appreciated. We also thank the contributors of photographs that were not in Marlin Rice's extensive collection. Individuals who provided illustrations for the identification and diagnostic keys and elsewhere in the handbook were Taina Litwak, Litwak Illustration Studio, Gaithersburg, MD; Jim and John Kalisch, University of Nebraska; and Mick Greenberg, Champaign, IL. Taina Litwak supplied all of the newly developed illustrations in the keys of this handbook. Several illustrations developed for the *Handbook of Soybean Insect Pests* by Jim and John Kalisch also were used in the keys. Mick Greenberg provided the graphic illustrations in "Introduction to Corn" and "European Corn Borer." Maps were developed by Media Graphics at Iowa State University. All of the illustrators did outstanding work, and we thank them for it. Their drawings significantly enhance the educational value of the handbook. Julie Todd, Iowa State University, read the entire manuscript and offered outstanding editorial assistance, ultimately strengthening the flow of text from one section to the next.

Many individuals at the Entomological Society of America provided assistance during production of this handbook. Most notably and most recently, Dawn Davern and Alan Kahan lent their professional expertise to the development of the handbook, and their help and support have been invaluable for completing production of the handbook. We offer our hearty thanks for their selfless efforts.

For the years that this handbook was under development, many secretaries at the editors' institutions helped with typing, packaging, and mailing, and we thank them all, particularly the secretaries at the University of Illinois and Iowa State University. And finally, we thank our families for supporting us during the long process of creating and completing this handbook.

The Editors
May 1999

Contributing Authors

John N. All
Department of Entomology
University of Georgia
Athens, GA 30602

David A. Andow
219 Hodson Hall
1980 Folwell
Department of Entomology
University of Minnesota
St. Paul, MN 55108

T. L. Archer
Texas Agricultural Experiment Station
Texas A&M University
Agricultural Research and Extension Center
Rt. 3, Box 219
Lubbock, TX 79401

Ramesh Ayyappath
Department of Entomology
Insectary, Tower & Judd Falls Road
Cornell University
Ithaca, NY 14853

J. E. Baker
USDA-ARS
U. S. Grain Marketing and Production Research Center
1515 College Avenue
Manhattan, KS 66502

Marlin K. Bergman
Pioneer Hi-Bred International, Inc.
7300 NW 62nd Avenue
P.O. Box 1004
Johnston, Iowa 50131–1004

Ricardo Bessin
Department of Entomology
S-225 Ag Science Building-North
University of Kentucky
Lexington, KY 40546

James W. Bing
Mycogen Seeds
301 Campus Drive
P.O. Box 280
Huxley, IA 50124

Larry W. Bledsoe
Department of Entomology
1158 Entomology Hall
Purdue University
West Lafayette, IN 47907–1158

Donald Bullock
Department of Crop Sciences
University of Illinois
W-201E Turner Hall
1102 South Goodwin Avenue
Urbana, IL 61801

Dennis D. Calvin
Department of Entomology
501 Agricultural Sciences and Industries Building
Pennsylvania State University
University Park, PA 16802

John L. Capinera
Department of Entomology and Nematology
Bldg. 970, P.O. Box 110620
University of Florida
Gainesville, FL 32611–0620

Moshe Coll
Department of Entomology
Faculty of Agricultural, Food and Environmental Quality Sciences
The Hebrew University of Jerusalem
P.O. Box 12, Rehovot 76100
Israel

Frank M. Davis
Crop Science Research Lab
USDA-ARS
P.O. Box 5367
Mississippi State, MS 39762

Richard A. Dolbeer
USDA-APHIS-ADC
National Wildlife Research Center
6100 Columbus Avenue
Sandusky, OH 44870

Bastiaan M. Drees
Department of Entomology
Texas A&M University
College Station, TX 77843–2475

Catherine E. Eastman
Center for Economic Entomology
Illinois Natural History Survey
172 Natural Resources Building
607 East Peabody Drive
Champaign, IL 61820

C. J. Eckenrode
Department of Entomology
New York State Agricultural Experiment Station
Geneva, NY 14456

C. Richard Edwards
1158 Entomology Hall
Department of Entomology
Purdue University
West Lafayette, IN 47907–1158

Wayne A. Gardner
Department of Entomology
1109 Experiment Station
Griffin, GA 30223

Michael E. Gray
Department of Crop Sciences
University of Illinois
S-320 Turner Hall
1102 South Goodwin Avenue
Urbana, IL 61801

Harry R. Gross (deceased)

Michael J. Haas
3307 Ellen Avenue
Lansing, MI 48910

John J. Hamm
USDA-ARS-IBPMRL
P.O. Box 748
Tifton, GA 31793–0748

Ronald B. Hammond
Department of Entomology
Ohio Agricultural Research & Development Center
The Ohio State University
Wooster, OH 44691

Gary L. Hein
University of Nebraska
Panhandle Research and Extension Center
4502 Avenue I
Scottsbluff, NE 69361

Randall A. Higgins
Department of Entomology
Waters Hall
Kansas State University
Manhattan, KS 66506

Michael J. Horak
Department of Agronomy
3701 Throckmorton Hall
Kansas State University
Manhattan, KS 66506–5501

Randy Hudson
Department of Entomology
University of Georgia
P.O. Box 1209
Tifton, GA 31793

Jan J. Jackson
Pioneer Hi-Bred International
Windfall Indiana Research Center
3820 North at 100 East
Windfall, IN 46076

L. Von Kaster
Garst Seed Company
2369 330 Street
P.O. Box 500
Slater, IA 50244

Armon J. Keaster
Department of Entomology
1-87 Agriculture Building
University of Missouri
Columbia, MO 65211

Robert W. Kieckhefer
3226 Sunnyview Drive
Brookings, SD 57006–4282

Contributing Authors

Allen Knutson
Texas A&M University
Texas Agricultural Extension Service
17360 Coit Road
Dallas, TX 75252–6599

Douglas A. Landis
Department of Entomology
104B Pesticide Research Center
Michigan State University
East Lansing, MI. 48824–1311

Eli Levine
Center for Economic Entomology
Illinois Natural History Survey
172 Natural Resources Building
607 East Peabody Drive
Champaign, IL 61820

Leslie C. Lewis
USDA-ARS, Corn Insects and Crop Genetics Research Unit
Genetics Laboratory, c/o Insectary Bldg.
Iowa State University
Ames, IA 50011–0001

Robert E. Lynch
USDA-ARS-IBPMRL
P.O. Box 748
Tifton, GA 31793–0748
Charles E. Mason
Department of Entomology
University of Delaware
Newark, DE 19717–1303

David A. McCartney
Ohio Agricultural Research & Development Center
Department of Entomology
1680 Madison Ave.
Wooster, OH 44691–4096

Murdick J. McLeod
Pioneer Hi-Bred International, Inc.
1000 West Jefferson
Tipton, IN 46072

G. J. Michels, Jr.
Texas A&M University
Research and Extension Center
P.O. Drawer 10
Bushland, TX 79012

Loren J. Moshier
Information Systems
Asgrow Seed Company
1212 George Drive
Normal, IL 61761

Gary P. Munkvold
Department of Plant Pathology
Iowa State University
317 Bessey Hall
Ames, IA 50011–1020

Emerson D. Nafziger
Department of Crop Sciences
University of Illinois
W-301 Turner Hall
1102 South Goodwin Avenue
Urbana, IL 61801

Lowell R. Nault
Administrative Service Building
Ohio Agricultural Research & Development Center
The Ohio State University
Wooster, OH 44691

Donald A. Nordlund
USDA, ARS, MSA
Biological Control and Mass Rearing Research Unit
PO Box 5367
Mississippi State, MS 39762

John J. Obrycki
Department of Entomology
Iowa State University
Ames, IA 50011–3140

Kenneth R. Ostlie
Department of Entomology
University of Minnesota
219 Hodson Hall
St. Paul, MN 55108

Frank B. Peairs
Department of Bioagricultural Sciences and Pest Management
Colorado State University
Fort Collins, CO 80523–1177

M. W. Phoofolo
Department of Biology
University of Lesotho
Lesotho
Africa

Clinton D. Pilcher
Monsanto Agronomy Center
1677 80th Street
Monmouth, IL 61462

Sharron S. Quisenberry
College of Agriculture
Montana State University
Bozeman, MT 59717

John C. Reese
Department of Entomology
Kansas State University
Waters Hall
Manhattan, KS 66506–4004

Marlin E. Rice
Department of Entomology
103 Insectary
Iowa State University
Ames, IA 50011–3140

Thomas J. Riley
Department of Entomology
402 Life Science Building
Louisiana State University
Baton Rouge, LA 70803

William B. Showers
64627 East Rosewood Drive
Tuscon, AZ 85739

Kevin L. Steffey
Department of Crop Sciences
University of Illinois
S-322 Turner Hall
1102 South Goodwin Avenue
Urbana, IL 61801

J. W. Stewart
HC 34–1125
Uvalde, TX 78801

Benjamin R. Stinner
Ohio Agricultural Research & Development Center
Department of Entomology
1680 Madison Avenue
Wooster, OH 44691–4096

Robert L. Stoltz
University of Idaho
Twin Falls Research & Extension Center
P.O. Box 1827
Twin Falls, ID 83303–1827

Gerald R. Sutter
CR Consulting, Inc.
2005 Iowa Street
Brookings, SD 57006

J. E. Throne
USDA-ARS
U. S. Grain Marketing and Production Research Center
1515 College Avenue
Manhattan, KS 66502

Jon J. Tollefson
Department of Entomology
17 Insectary Building
Iowa State University
Ames, IA 50011–3041

Noel N. Troxclair, Jr.
Texas A&M University
P. O. Box 1849
Uvalde, TX 78802–1849

John W. Van Duyn
Department of Entomology
North Carolina State University
Vernon G. James Research and Extension Center
207 Research Station Road
Plymouth, NC 27962

D. R. Webb
2917 Orleans Seneca Road
Clifton Springs, NY 14432

John L. Wedberg
Department of Entomology
237 Russell Labs
University of Wisconsin
Madison, WI 53706

Michael J. Weiss
Department of Plant, Soil and Entomological Sciences
University of Idaho
Moscow, ID 83844

Robert N. Wiedenmann
Center for Economic Entomology
Illinois Natural History Survey
172 Natural Resources Building
607 East Peabody Drive
Champaign, IL 61820

Contributing Authors

Gerald E. Wilde
Department of Entomology
Kansas State University
Manhattan, KS 66506

Harold R. Willson
IPM Program
The Ohio State University
1991 Kenny Road
Columbus, OH 43210

B. R. Wiseman
214 Fulwood Blvd.
Tifton, GA 31794

John F. Witkowski
Northeast Research & Extension Center
University of Nebraska
Box 111
Concord, NE 68728–0111

Robert J. Wright
Department of Entomology
South Central Research and Extension Center
University of Nebraska
P.O. Box 66
Clay Center, NE 68933–0066

Roger R. Youngman
Department of Entomology
216 Price Hall
Virginia Polytechnic Institute & State University
Blacksburg, VA 24061–0319

How to Use This Handbook

By Kevin L. Steffey

This handbook is designed primarily for practitioners of integrated pest management (IPM) programs in corn, including growers, crop consultants, extension agents, and company agronomists and sales representatives. However, students and researchers also will find valuable information in this handbook. Our primary objective is to provide fundamental and useful information about management of corn insects throughout the United States and Canada. Although this handbook focuses on insect pests of corn, noninsect pests (e.g., mites, symphylans, and slugs) also are covered. The first three sections of the handbook provide information about corn and corn production, principles and practices of corn insect management, and an overview of the pests that occur in corn, including insects, weeds, plant pathogens, birds, and mammals. The remainder of the handbook is devoted to discussions of insects and related pests of corn and to beneficial organisms. Authors are identified at the end of each section, and a list of references for additional information is provided for most sections. A glossary provides definitions for unfamiliar words. A section on sources of local information provides addresses of entomology departments throughout the United States and Canada.

Identification of Insects and Diagnosis of Injury

The introduction to identification of insects and diagnosis of injury (page 22) explains how to use the list of corn insect pests grouped by injury classification, the key to corn injury caused by insect pests, and the key to corn pests. The list and keys provide directions to sections in the handbook in which color photographs of the insects and examples of injury to corn accompany detailed discussions about the pests. The keys and the color photographs provide assistance for identifying insects and diagnosing injury caused by insects in corn. Unknown insects and unfamiliar symptoms of injury often can be identified first by using the keys and then by comparing specimens with the color photographs.

Insect Pests of Corn and Their Management

Insect pests that attack corn are discussed in sections arranged alphabetically by common name for easy reference. Within each section, information is provided on scientific classification, origin and distribution, description (aids in identification), pest status (relative importance of the pest), injury (aids in diagnosis), life history, and management.

Suggestions for management are included as an overview of control tactics that are effective for a specific pest. If insecticides are useful for managing a pest, they are mentioned, but trade names are not identified. Economic thresholds, when known, also are provided. However, economic thresholds may vary from one region of North America to another. Refer to a state's Cooperative Extension Service guidelines for management of insects in corn and to obtain information about specific insecticides and regionally accepted economic thresholds.

Beneficial Organisms

The section on beneficial organisms, including insect pathogens (entomopathogens), parasitoids, and predators, provides information about their potential for natural or applied biological control of corn insect pests. The information varies among organisms because of their diversity. However, a genuine attempt was made to include all or most of the following information for each organism: scientific classification, origin and distribution, description, biology or life history, degree of natural control, sampling methods (where appropriate), potential for biological control, and effects of crop or insect management on the organism. Tables in some sections list the corn insect pests infected (pathogens), parasitized, or preyed upon by the different beneficial organisms.

An Introduction to Corn

By Emerson D. Nafziger and Don Bullock

History of Corn as a Crop

Corn (*Zea mays*) is, perhaps more than any other crop of worldwide importance, a product of human selection and improvement. The corn crop was domesticated more than 7,000 years ago in Central America, and the domestication process was dramatic and ingenious. The development of most food crops originated with wild ancestors that were selected over a long time to produce more food of higher quality. Evidence suggests that the domestication of corn (probably from a wild grass known as teosinte) was so complete that corn cannot exist in the wild and few, if any, wild relatives still exist. From its center of origin, corn (called maize throughout most of the world) spread to become the major food staple of most people living in the Americas in pre-Columbian times.

When Europeans came to the Western Hemisphere, they found corn growing from the Caribbean Islands to New England. Several different types of corn were being produced, from flour corns with very soft starch in their kernels to the hard-kernel "flint" corns. Great Plains and Eastern Woodland dwellers were dependent upon corn, seed of which likely had reached the interior of North America by way of trade with people living in the Southwest, including modern-day Mexico.

The development of modern Corn Belt hybrids, which have made corn such a successful crop, began with a genetic "accident" several hundred years ago. The European settlers in the southeastern United States discovered a type of corn growing there that was different from the short-growing flint corn common in the northeastern United States. The plants were taller, the ears were larger, and the kernels generally were dented on their outer end. When the two types of corn were grown near each other, some of the plants resulting from cross-pollination had good traits from both types of corn. They produced larger yields than the flint corn but had some of the flints' good kernel characteristics. Farmers in the middle Atlantic region continued to grow these superior types.

As settlers moved west during the early 1800s, they carried these superior corn varieties with them to plant on the newly opened land. Fields of such varieties consisted of many plants that varied in maturity, plant height, and kernel traits. Innovative farmers learned that if they kept ears from the better plants for seed, the next generation of corn plants retained some of the better characteristics, and consequently, produced higher yields. Some of these varieties became known for their originators (Reid's Yellow Dent).

At the turn of the century, plant scientists confirmed genetic principles that revolutionized plant breeding. Early studies showed that when two different individual plants were crossed, their progeny usually performed better than either parent, a trait known as "hybrid vigor." Crosses between individuals from different corn varieties expressed hybrid vigor, and such hybrids produced much higher yields compared with inbreds. The first commercial hybrid seed corn was grown in Iowa in 1923. The cross was forced by hand-removal of the tassel (male flower) on the seed plants, so that pollen to fertilize the kernels on those plants had to come from the second variety grown in the same field. Although this method of seed production was labor intensive and yields of the seed parents were not always good, the production of hybrids spread rapidly in the 1930s and 1940s. Today, virtually all corn produced in the United States is grown from hybrid seed. Breeding efforts are directed toward improvement of populations (such as in the early varieties), from which superior inbreds are selected to serve as hybrid parents.

Importance of Corn in the World

Although corn originated in a subtropical climate, it has adapted to many climates around the world, from the lowland tropics to cool areas more than 7,000 ft (2,100 m) above sea level. Much of the corn grown in other parts of the world, especially in temperate climates such as in western Europe, originated from U. S. Corn Belt hybrids, with genetic improvement to adapt it to local conditions.

Corn is the most important "coarse grain" crop (i.e., a crop used mostly as livestock feed rather than as human food) in the world. In 1994, a record year for corn production, 555 million metric tons (MMT) of corn was produced worldwide; the United States produced 257 MMT, or about 46% of the total. Other than the United States, only China, with 99 MMT, produced more than 10% of the world's corn in 1994. Major corn-producing countries in Europe are France, Italy, Spain, Germany, the former Yugoslavia, and Romania.

Although corn is grown throughout the United States, the largest concentration of corn acreage is in the region from Nebraska to Ohio, commonly called the Corn Belt (Fig. 1). As much as one-half of all crop acreage in the Corn Belt is planted to corn each year. Yields vary considerably from year to year, but with the excellent growing conditions in 1994, the average yield in the United States was 139 bushels per acre (8.7 tons per hectare) on 72.9 million acres (29.5 million hectares).

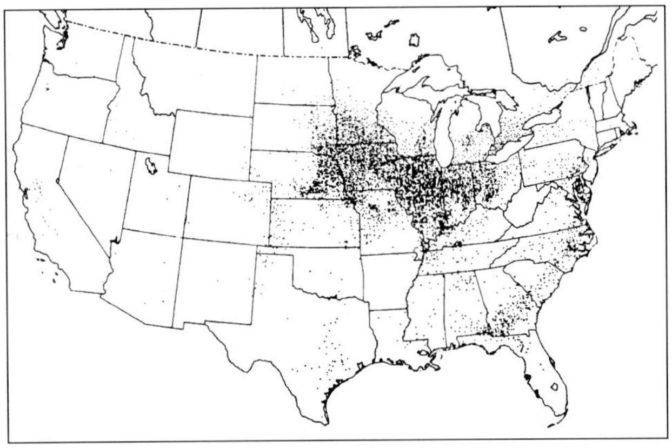

Fig. 1. Map of the United States and Canada showing acreage of corn in 1994. Each dot represents 10,000 acres (4,050 hectares).

Although most of the world's corn is used as animal feed, corn grown for human consumption is important in some regions. Ground corn is a staple food in much of Africa, and coarse flour from corn is formed into tortillas, a food staple in parts of Latin America. In North America, corn is consumed as snack food, breakfast cereal, and cornmeal grits and cakes. High-fructose corn sweeteners, cornstarch-based food ingredients, corn oil, and other products also are important. Some commercial breeding efforts currently are focused upon producing hybrids with grain characteristics that make them more suitable for specific processing uses.

Corn Plant Development

Corn is a monocot that originated in a warm area, so its growth rate is responsive to warm temperatures. The growth rate of corn is linear over the range of average daily temperatures from 50°F (10°C) to 86°F (30°C). Consequently, growth rates based upon accumulated daily temperatures that are measured in units called Growing Degree Days (GDD) can be predicted and monitored. Growing Degree Days are calculated as follows:

$$\text{GDD} = \frac{(\text{minimum daily temperature} + \text{maximum daily temperature})}{2} - 50^\circ\text{F}$$

The base temperature, 50°F, is subtracted because corn grows very little at or below this temperature. A minimum temperature of less than 50°F should be recorded as 50°F to calculate GDD. Because growth rates do not increase at temperatures higher than 86°F, a maximum temperature of more than 86°F should be recorded as 86°F to calculate GDD. Thus, the minimum GDD accumulation for a day is zero (average daily temperature less than or equal to 50°F), and the maximum is 36 GDD, which would accumulate only if the minimum temperature were 86°F or higher. Twenty-five GDD would accumulate on a typical summer day with a low temperature of 65°F (18.3°C) and a high temperature of 85°F (29.5°C).

Growing degree days can be used to classify hybrid maturity and predict plant developmental stages. Growing degree-day requirements for hybrids grown in the Corn Belt range from about 2,200 GDD for early hybrids to about 3,000 GDD for late-maturing hybrids grown in the southern part of the region. A mid-season hybrid adapted to central Illinois might require about 2,700 GDD from planting to maturity. The following table provides approximate GDD accumulations required to attain different growth stages for a hybrid of this maturity:

Stage	GDD
Emergence	120
2-leaf	200
6-leaf	475
10-leaf	740
14-leaf	1,000
Tassel emergence	1,250
Silking	1,400
Kernel dough stage	1,925
Kernel dent stage	2,450
Physiological maturity	2,700

The total GDD required for maturity varies among hybrids, and the relative proportions of full-season GDD requirements to reach certain growth stages change only slightly with maturity. About 65 GDD are required to produce each new leaf during vegatation de-

Fig. 2. Germination and emergence of corn 0, 2, 4, 5, 6, 7 (VE), 8, 10 (V1), and 12 (V2) days after planting (Iowa State University).

velopment. Early season hybrids require fewer leaves, so the planting-to-silking period is shorter than the silking-to-maturity period.

Although several systems are used to describe the growth stages of corn, a commonly accepted system, developed at Iowa State University, describes the VE (emergence) stage, followed by successive V*n* (vegetative) stages, where *n* is the number of leaves with the collar visible. The germination and emergence of a corn seedling are illustrated in Fig. 2. After germination, the coleoptile of the plant grows toward the soil surface. After the tip of the coleoptile reaches the light, the formation of the crown begins at a depth of about 1 in (2.5 cm) in the soil. Thus, depth of seeding does not affect depth of the crown very much, but the mesocotyl will be longer the more deeply seeds are planted.

At about the V6 (6-leaf) stage, leaf initiation at the tip of the stem stops, and the tassel is initiated. The growing point is located at or slightly above the soil surface as elongation of the lower nodes of the stem begins (Fig. 3). Stem elongation accelerates considerably after this stage, resulting in a rapid increase in plant height. Rapid growth continues until tassel emergence. Ear shoots are initiated on nodes above ground at the V6–V7 stages. The uppermost ear, which usually is the only one on the plant that develops fully, becomes the predominant ear by about stage V12.

Tassels emerge from the leaf whorls at about stage V19, or about the time the last leaf (usually leaf 20 or 21) is near full emergence. The ear shoot becomes visible as it emerges from the sixth or seventh leaf sheath below the top of the plant. Full emergence of the tassel is called the VT stage (Fig. 4) and signals the end of vegetative stages, even though some increase in size of the vegetative portion of the plant continues for a week or so longer. The root system, consisting almost entirely of adventitious roots arising from the lower six or seven nodes of the stem, also reaches a maximum size at about VT, and may extend to a depth of 4–5 ft (120–125 cm). However, most of the active roots are in the upper 6 in (15 cm) of the soil.

Fig. 3. V6 plant (Iowa State University).

Fig. 4. VT plant (Iowa State University).

Within a few days after tassel emergence, pollen is shed by anthers attached to tassel branches. With adequate soil moisture and favorable temperatures, silks attached to kernels at the base of the ear begin to appear a day or so after pollen shed begins. Silks grow mostly at night because water is more available in the plant, and most pollen is shed as the relative humidity decreases in the morning. Newly emerged silks catch pollen grains as they fall. Pollen germinates and grows toward the kernel where fertilization takes place. New silks are exposed each day for 3–4 days; pollen shed usually continues for 1–2 days after all silks have emerged.

Drought, loss of leaf area, and injury to silks can have devastating effects on the pollination process. During severe drought, silks may fail to emerge, resulting in complete failure of fertilization and kernel development. Very high temperatures (90°F [32.2°C] or higher) decrease pollen viability and perhaps decrease receptivity of the silks to pollen. Factors that decrease effective leaf area (e.g., hail, disease, insects) also lower the number of kernels that are fertilized by decreasing the carbohydrate supply to the reproductive parts of the plant. Several insects affect pollination by eating silks, thus decreasing the chances for successful capture and growth of pollen. Finally, kernels may abort after fertilization, usually as a result of continued drought.

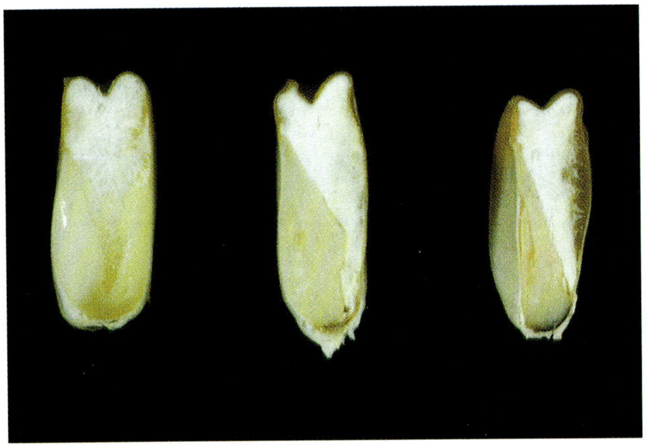

Fig. 5. Black layer development (Iowa State University).

Beginning with pollination, plant developmental stage is designated according to kernel (reproductive) development, from R1 (silking) through R6 (physiological maturity). Embryo development in the kernel continues through the "blister stage" (R2) and "milk stage" (R3). Starch deposition, beginning with the "dough stage" (R4) and through the "dent stage (R5)," continues until sugar movement into the kernel stops and an abscission layer ("black layer") forms at the base of the kernel (Fig. 5). Maintenance of a healthy leaf canopy is critical during the grain-filling stage; any factor that decreases leaf area decreases grain yield by limiting the supply of photosynthate available to fill the kernels.

Climatic Requirements for Corn

Shortage of water is the most important yield-limiting factor in most areas where corn is grown. Corn requires about 22 in (560 mm) of water to produce a high yield, although water use by corn varies from about 18–25 in (460–635 mm), depending upon evaporative demand. Corn is one of the most water-consumptive crops. Corn uses water most during June through August in the Corn Belt; peak water use of as much as 0.2 in (5 mm) per day occurs during mid-July. The corn plant is most sensitive to drought during pollination in mid-July, which also is the time of year when the frequency of drought in the Midwest is greatest.

Average rainfall in July in the Corn Belt usually is not sufficient to meet the needs of the corn crop. In dry years, the deficit between the crop requirement and rainfall amount may be as much as 4 in (100 mm) during July, which normally would be disastrous for corn. However, many of the more productive soils hold as much as 10 in (250 mm) of plant-available water in the rooting zone. The Corn Belt is well suited for corn production because of the depth and excellent physical texture of its soils. Enough water is stored in the rooting zone in deep silt loam or silty clay loam soils to provide as much as half of the crop requirement for the growing season. In sandy soils or in soils where root growth is restricted (compacted or poorly drained soils), the amount of available water may be much less than ideal, so short periods of drought may decrease yield significantly.

Temperatures for corn production in the Corn Belt usually are more ideal than is water supply. Corn thrives when temperatures are warm but not when temperatures are very hot. The ideal temperature during the day is 85–90°F (30–32°C). The ideal temperature at night is 60–65°F (15–17°C). Corn grows best when temperatures in June are about normal, when temperatures in July are well below normal, and when temperatures in August are slightly above normal. Temperatures above 95°F (35°C) usually do not harm corn unless soil moisture is also low. Temperatures below 50°F (10°C) can cause temporary harm to the corn plant; serious loss of leaf function occurs if temperatures decrease to about 40°F (4°C).

Sunlight is another climatic element important to corn. Relative to other crops, corn converts sunlight to dry matter very efficiently, even though the efficiency of conversion is only a small percentage. Corn thrives on bright sunlight, although the close association between bright sunlight and drought may obscure the advantages of sunlight. If enough water is available to the crop, however, more sunlight probably produces higher yields.

Corn Management

Hybrid Choice. More effort has been expended on the genetic improvement of corn than on any other crop, resulting in many good hybrids from which a grower can choose. Many companies with large corn breeding programs produce their own hybrids; other companies obtain their new hybrids from breeding programs maintained by "foundation" seed companies. Regardless of the source, most corn hybrids on the market perform reasonably well under most conditions.

A corn grower faces a bewildering array of choices of hybrids and some aggressive efforts by companies to sell their hybrids. Sources of information about corn hybrids include demonstration field comparisons, company literature, and results from corn performance trials conducted by personnel at public universities in most states. Although the ability of a hybrid to produce high yields is the most important criterion for corn growers, the ability of stalks to remain standing until harvest and a corn hybrid's tolerance to diseases and insects are also selection criteria.

The development of hybrid corn seed was the first in a number of technological advances that increased corn yields dramatically from their rather static levels in the 1940s. Although corn yields have increased primarily because of genetic improvement, the development and use of nitrogen fertilizer and pesticides also have contributed significantly to improving yields in recent years. Controlling weeds and providing plant nutrients as needed allows flexibility that enables farmers to change their production practices in response to the need to reduce input costs per bushel.

Soil Fertility. Corn is a high-yielding crop, so the requirements for nutrients and water are great. Corn yields best when planted in a deep, well-drained soil with high water-holding capacity and adequate levels of plant-available nutrients. In addition to nitrogen (N), nutrients that must be supplied to corn include phosphorus (P), potassium (K), magnesium (Mg), calcium (Ca), sulfur (S), and some "micronutrients" that are required in very small quantities. Because the availability of most of these nutrients is affected by the degree of acidity of the soil, the pH of the soil also must be within an acceptable range, ideally pH 6–7. Nitrogen fertilizer lowers the pH, so limestone must be added periodically to maintain desirable pH levels.

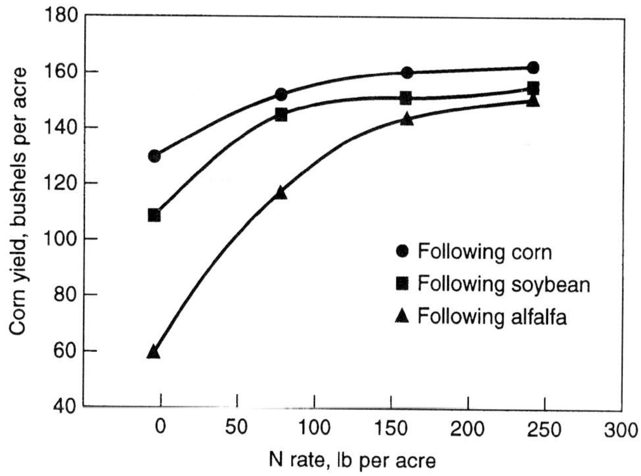

Fig. 6. The effect of previous crop on the response of corn to nitrogen. Data are from a study conducted at DeKalb, Illinois, 1980–1983.

Nitrogen is the nutrient needed in greatest quantities for corn production, and N fertilizer is a costly input. In recent years, management of N fertilizers to prevent nitrate contamination of surface and groundwater supplies has become more important. Results from numerous experiments have identified optimum rates of N fertilizer for corn grown in different soils in different cropping systems. Requirements for N fertilizer are greatest when corn is planted after corn and lower when corn is planted after a legume crop such as soybeans or alfalfa (Fig. 6). Although many studies regarding rates of N have been conducted, recommendations for N fertilizer usually do not indicate absolute rates to be used; rather, rates of N to be applied are based upon expected yield in a particular field. For example, the current recommendation in Illinois is to supply about 1.2 lb of N per bushel (19.5 kg N per ton) of expected yield. The amount of N to be supplied by fertilizer may be decreased if the previous crop was a legume (e.g., recommended rates are decreased by 40 lb N per acre [45 kg per hectare] if corn is planted after soybeans in Illinois), livestock manure (which contains N) was applied, or a soil test in the spring reveals that the soil contains a substantial amount of plant-available N.

Tillage and Crop Rotation. Corn growers have reduced the amount of tillage done in preparation for planting to improve soil conservation and reduce production costs. The reduction in the amount of tillage ranges from using a chisel plow instead of a moldboard plow, thereby increasing the amount of surface residue left from the previous crop, to planting corn without any previous tillage (no-till). Yield generally has not been affected much by switching from the moldboard plow to the chisel plow, but the effect of no-till production on yields of corn varies. In lighter, better-drained soils, especially in warmer areas of the United States, no-till practices have resulted in corn yields equal to or greater than yields achieved in more conventional tillage systems. In heavier soils, especially in northern areas, no-till practices usually result in placement of seeds into cooler, wetter soil, which sometimes results in slower growth, decreased plant populations, and perhaps lower yields. Planting no-till corn into corn residue also may result in better survival of certain insects and disease pathogens. Without tillage to destroy weeds, weed control may be more costly and difficult in no-till systems than in conventional tillage systems.

Most of the corn in the Corn Belt is planted in an annual rotation with soybeans, a rotation that results in higher yields of both crops than if either crop were planted without rotation. The yield benefit from this annual rotation likely is due both to a combination of soil factors (such as N and P supply) and a decrease in the incidence of disease, insect, and weed infestations. Where forage legumes are grown for livestock, crop rotations usually are 3–5 years in duration, often including a small grain (wheat, oats) and 2–3 years of a persistent forage crop. However, continuous corn production is still practiced on many acres. Although evidence indicates that some of the problems associated with a short-term corn-after-corn sequence diminish over time, continuous corn seldom yields as well as corn planted after another crop.

Planting. Optimum planting time for corn in the central Corn Belt is during the last half of April; little yield loss occurs if the corn can be planted by early May (Fig. 7). Yields usually decrease if planting is delayed beyond early May. Late planting delays pollination until a hotter, drier time during the summer, and temperatures late in the grain-filling period usually are too cool for grain maturity. Diseases and insects cause more problems in late-planted corn at an earlier stage in the grain-filling process when injury usually causes greater yield decreases. Disadvantages associated with planting too early include the need for larger equipment to avoid weather delays (early planting must be accomplished quickly) and poor weed control in cool soils.

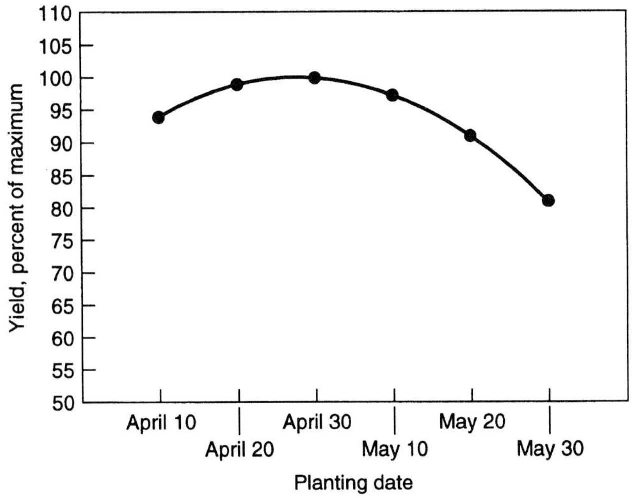

Fig. 7. Corn yield response to planting date, at a final plant population of 30,000 plants per acre. Data are from a study conducted at Monmouth and DeKalb, Illinois, 1987–1990.

Although planting date often is determined by weather conditions, growers can manipulate corn plant populations easily. Reported plant populations at harvest time generally have been less than 24,000 plants per acre in most Corn Belt states; however, some producers probably could increase their yields if they planted more seeds (Fig. 8). In productive Corn Belt soils, the optimum plant population may be 25,000–30,000 plants per acre at harvest time. At the normal establishment rate of about 90% of the seeds planted, 28,000–33,000 seeds per acre are required to obtain 25,000–30,000 plants per acre for maximum yields. In soils subject to drought, plant populations should be lower. However, modern hybrids are bred to pro-

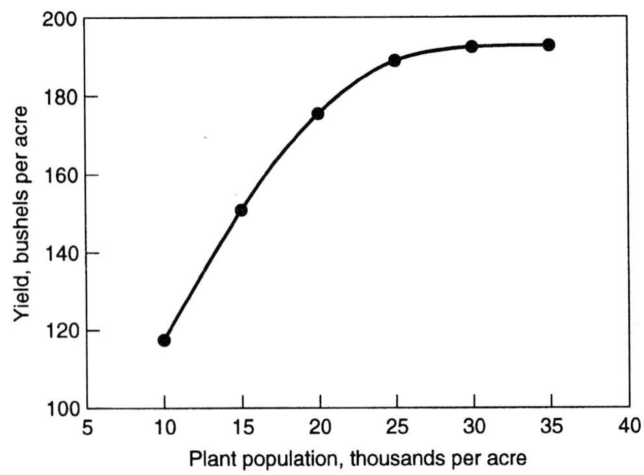

Fig. 8. Corn yield as affected by plant population. Data are the averages of six hybrids grown in studies conducted at Urbana, Monmouth, and DeKalb, Illinois, 1991–1994.

duce well even when the environment stresses the plants, so final plant populations in soils subject to drought should not be much lower than 20,000–22,000 plants per acre.

The average distance between rows has decreased from about 40 in (102 cm) when horses provided most of the power on farms to 30 in (76 cm) on most farms today. Although the decrease in row spacing probably has increased yields in most instances, the effect of decreasing row spacing below 30 in (76 cm) is less clear. The main improvement in yield when corn is planted in narrow rows likely derives from an increase in the amount of sunlight intercepted by the canopy during grain fill. Reducing the row spacing to less than 30 in (76 cm) may have little effect on yield if corn is planted at higher populations (Fig. 9) or if the hybrid produces large plants. Although the cost of equipment to plant and harvest narrower rows may be a disadvantage, the possibility that narrower rows may increase yields, particularly during years when plants do not grow well because of unfavorable weather conditions, has encouraged some farmers to invest in the equipment to plant and harvest corn grown in 15- to 20-in (38- to 51-cm) rows.

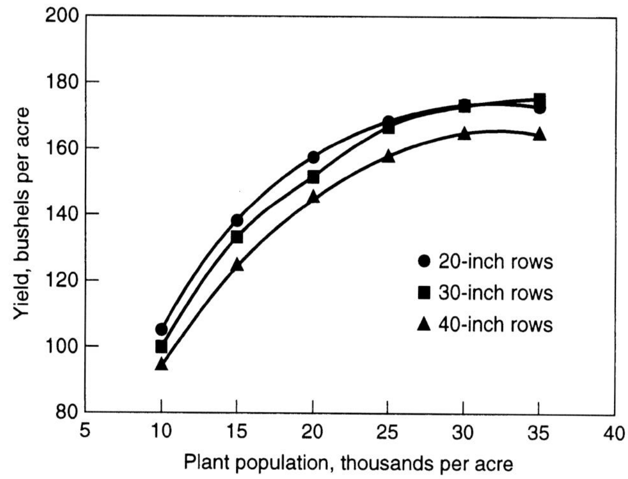

Fig. 9. Corn yield response to plant population at different row spacings. Data are from a study conducted at Monmouth and DeKalb, Illinois, 1992–1994.

Harvest and Storage. Corn reaches physiological maturity (black layer) at 30–32% grain moisture. Machine harvest can begin with most harvesting equipment when grain moisture is about 27% if the equipment is adjusted to prevent kernel damage. As moisture decreases below 20%, yield losses increase due to dropped ears and shelling at the corn head of the combine. Consequently, most machine harvesting should be completed before grain moisture decreases below 20%, unless certain circumstances (e.g., special uses for the corn) require more drying in the field. Harvested grain usually is dried with fossil fuel-powered driers. For safe storage, corn grain must be dried to 14–15% moisture. When stored for a long time, corn also should be aerated periodically to prevent moisture migration caused by uneven temperatures in the storage structure.

Selected References. 129, 213, 232

Corn Insect Management

By Robert J. Wright and John W. Van Duyn

Introduction and Principles

Integrated pest management is a concept proposed for use by growers to prevent or limit pest damage by planning and making decisions that balance the positive and negative aspects of the tactics used to limit pest damage. In application, IPM systems are tailored for specific situations, such as groups of crop fields or whole farms, and should embrace all considerations of the pest management tactics used, including, but not limited to, effectiveness, cost, risks, and potential for environmental contamination.

The IPM concept originally focused upon insect management, emphasizing a reduction in the use of insecticides by relying first upon biological control, cultural practices, and other nonchemical tactics for pest management and by basing the application of insecticides on need. Later, IPM was expanded to include weeds, plant pathogens, and nematodes as well as the integration of pest management practices for different types of pests. Successful IPM programs allow producers to maintain profitable crop production and minimize the undesirable effects of pest management practices. This section describes the principles of IPM applied to insects in field corn.

Integrated pest management programs typically are based upon four fundamental strategies: *prediction, prevention, detection,* and *remedial action,* if necessary. These strategies provide a systematic approach for implementing IPM.

Prediction of pest occurrence involves anticipation of the kinds of pests that might develop within a field or on the farm and the likelihood that pest densities will reach or exceed economically damaging levels. Prediction of pest occurrence can be based upon past experience; counts of insects from the previous season; knowledge of pest biology, weather patterns, and cropping circumstances (e.g., no-till, crop rotation); and computer models that generate projections of insect pest populations. Management plans can be developed for the pests that are most likely to cause crop loss. The management plan should be appropriate for all pests, their potential to cause economic damage, and their likelihood of occurrence. Pest management tactics range from application of a costly preventive insecticide treatment for an insect expected to cause serious economic losses to monitoring during the growing season for pests that cause economic damage occasionally. Doing nothing when the pest density is below an economically damaging level is a legitimate pest management tactic.

Prevention enables a grower to avoid the pest or the injury it causes. Prevention may be accomplished completely or partially by moving the crop (crop rotation), planting an insect-resistant variety, increasing the tolerance of the crop to the pest with good agronomic practices, planting a hybrid with a different maturity, or otherwise creating a situation where the crop escapes the pest or its damage. Nonchemical preventive tactics for pest management are encouraged because they often are cheap, have few negative consequences, and are compatible with farming practices.

Detection is the process of discovering pests in a field or on the farm and estimating their potential to cause economic damage. Scouting and monitoring for pests enable a grower to observe the seasonal dynamics of pest populations. Economic thresholds have been developed to indicate when the density of a pest is large enough to cause economic damage to the crop and, therefore, justify a control action (e.g., an insecticide application).

Remedial action is taken when detection reveals that an economic pest problem exists. The intent of remedial action is the reduction of a pest density to a level below the economic threshold. Insecticide application currently is the most commonly used remedial tactic.

Growers should consider the consequences of selecting certain IPM tactics. Some considerations are obvious, such as availability, cost, effectiveness, and ease of use of the selected tactic. However, other important considerations include environmental safety, potential effect on human health, impact on beneficial insects, worker protection issues, legal limitations, and social consequences. All important factors should be evaluated to determine the best IPM plan for both short- and long-term impacts.

Integrated pest management must be an integral component of crop management. Management of the corn crop affects pest populations and crop tolerance to insect feeding. Growers likely will adopt pest management practices only if they are compatible with the goals and limitations of crop production.

Tactics for Corn Insect Management

Tactics for managing corn insects include cultural, biological, and chemical controls, and plant resistance. Specific management tactics are discussed in the sections about individual pests.

Cultural Control. Cultural control tactics for corn insects involve many of the common practices that must be used to produce and manage the crop, such as crop rotation, planting date, hybrid selection, plant population, soil fertility, irrigation, cultivation, and harvest date. These practices may prevent pests from occurring, reduce insect pest densities, reduce the impact of insect injury by improving the ability of the corn plants to recover from or tolerate insect injury, or reduce the length of time the crop is in a stage susceptible to insect injury. Crop rotation, early planting, establishment of a satisfactory plant population, rapid seedling growth, and early harvest are fundamental IPM tactics that reduce the potential for and severity of injury by some important insect pests in field corn.

Cultural control tactics should take advantage of an insect's biology. For example, in most areas of the United States, western corn rootworm females lay eggs in late summer only in cornfields, and the larvae that hatch the next year survive only on corn roots. Consequently, interrupting the rootworm's life cycle by annually rotating corn with a nonhost crop such as soybeans provides excellent control of this key pest. Continuous corn culture provides a suitable host for rootworms every year.

Natural and Applied Biological Control. Natural control of insect pests in corn occurs without human intervention. Many species of predators, parasitoids, and pathogens regulate populations of insect pests in corn, and some of these natural enemies suppress densities of insect pests to levels below economic thresholds. More details about biological control and beneficial organisms are provided in the section that begins on page 120.

Applied biological control is the selective and planned use of beneficial organisms for management of pests. Three major strategies of biological control are *conservation* of existing natural enemies, *augmentation* of natural enemies by periodic releases when they are needed, and *importation* and establishment of new species of natural enemies. Currently, conservation of natural populations of beneficial organisms and importation of natural enemies are the strategies used most often in corn IPM programs.

Conservation. Some crop production and pest management practices increase the potential for survival of beneficial organisms in or near cornfields. Studies have shown that conservation of crop residue and reduced tillage favor the survival of several ground-dwelling predators, including ground beetles and spiders, that may feed on corn pests. In contrast, the use of broad-spectrum insecticides kills many beneficial insects; those that survive may starve from lack of prey. Consequently, application of the lowest effective rate of an insecticide only when the pest density exceeds the economic threshold greatly reduces the negative effects of broad-spectrum insecticides. Where feasible and effective, the use of selective insecticides, e.g., products that contain *Bacillus thuringiensis* (*Bt*), improves survival of natural enemies. Resistant or tolerant corn hybrids also reduce the need for insecticide application.

Diverse cropping systems and noncultivated land adjacent to cornfields may provide overwintering sites and additional food sources (nectar, pollen, alternate prey) for beneficial insects. When economic densities of pests are not present, predatory and parasitic insects sometimes maintain their populations by surviving on noneconomic densities of the pests or on alternate prey on noncultivated plants or weeds in the field or in field margins. Beneficial insects influence populations of some pests in almost all fields each year. Their contribution in suppressing pest populations often is not realized.

Augmentation. Commercial-scale augmentation of biological control organisms in cornfields is not common because of the relatively low profit potential of field corn. Research is ongoing to evaluate the use of *Trichogramma* species (tiny wasps that parasitize insect eggs) for control of European corn borers and corn earworms. *Trichogramma* wasps likely will be used first in seed corn production because of the higher profit potential. Augmentation of natural enemies in corn is more common in countries where labor costs are low.

Importation. A few of our most important insect pests of corn (e.g., European corn borer) are not native to North America. These pests usually were introduced accidentally without their natural enemies that existed in their original range. Importing natural enemies and establishing them in new areas has improved biological control of several introduced pests. Importation usually has been conducted by federal and state government agencies (United States Department of Agriculture [USDA], state departments of agriculture). However, before a new natural enemy can be released, researchers must determine that the natural enemy is adapted to the target area and, thus, likely will become established; that it has a limited range of prey and does not harm beneficial insects; and that it does not harbor its own natural enemies (diseases, parasitoids) that could attack other species of natural enemies if it were released. The USDA-APHIS (Animal and Plant Health Inspection Service) has imported and released the parasitic fly *Lydella thompsoni* to try to establish it in areas where European corn borers are a major pest.

Chemical Control and Insecticides. Insecticides are the most commonly used remedial control tools for insects attacking corn. Insecticides are relatively easy to use, reduce pest densities immediately, and enable a grower to respond quickly when pest densities increase above economic thresholds. Disadvantages of chemical insecticides include the potential for detrimental effects to human health, the environment, and the ecosystem, particularly if they are not used properly. The most commonly used corn insecticides are organophosphates, carbamates, and pyrethroids. Pheromones and insecticidal baits also can be categorized as chemical control tactics. Although they are not chemicals, microbial insecticides (primarily products containing *Bt* protein toxins) are used remedially in much the same manner as chemical insecticides.

Selective insecticides and selective use of insecticides minimize the adverse effects of insecticides on nontarget organisms, including beneficial insects (predators, parasitoids, and honey bees), wildlife, and humans. Selectivity of an insecticide depends upon its mode of action and its formulation. For example, some insecticides can be delivered in a bait formulation that greatly reduces the amount of chemical required for effective control. Selective use of insecticides can be achieved by the timing, rate, and method of application, and the site to which the insecticide is applied. Efficient use of insecticides requires selection of the appropriate insecticide for the pest to be controlled and precise application. Proper calibration and operation of application equipment are essential for effective and economic use of insecticides.

Continuous use of some insecticides has resulted in the development of insecticide resistance in some species of corn insect pests. Resistance to cyclodiene insecticides was detected in 1959 in western corn rootworms in south-central Nebraska; the trait subsequently spread among populations of this pest throughout the Corn Belt. More recently, resistance to the aerially applied insecticides methyl parathion and carbaryl has been detected in populations of western corn rootworms in Nebraska.

Management of insecticide resistance is an important component of IPM. Resistance management strategies will maintain the availability and effectiveness of corn insecticides. A combination of applying insecticides only when necessary (adherence to economic thresholds) and at the minimal effective rates; rotating among different classes of insecticides, particularly those with different modes of action; and using appropriate nonchemical control tactics reduces the speed at which insecticide resistance develops. Spot treatments within a field and treatments targeted to control only a portion of a pest's population also are resistance management strategies.

Plant Resistance to Insects. Corn hybrids vary in their susceptibility to insect pests and their suitability as food sources for insects. Plant breeding programs have incorporated resistance to a few major insect pests into commercial hybrids. For example, most commercial corn hybrids have some degree of resistance to whorl-stage feeding by European corn borer larvae, and the level of resistance often is listed in seed corn catalogs. Resistance to corn borers is imparted by the presence of a plant aglucone, 2-4 dihydroxy-7-methoxy-1, 4-benzoxazin-3-one (DIMBOA) that is most concentrated in whorl-stage corn.

In recent years, insect-resistant corn hybrids have been developed through transgenic biotechnology. Modern gene transfer techniques have been used to develop corn plants that contain a gene taken from the naturally occurring soilborne bacterium *Bacillus thuringiensis* (*Bt*). The gene produces a crystalline protein (endotoxin) that is toxic to specific groups of insects. In corn, the endotoxin is specific for control of certain caterpillar pests. Current commercially available *Bt* corn hybrids are resistant to some of the stalk-boring caterpillars, e.g., European corn borer, southwestern corn borer, and southern cornstalk borer. Some *Bt* corn hybrids also reduce the amount of insect injury in the ear, thus decreasing the level of mycotoxins (e.g., aflatoxins) in the grain. Several different *Bt* proteins and resistance factors other than *Bt* endotoxins are being studied to determine their potential for use in transgenic corn hybrids to control other corn insect pests (e.g., corn rootworms).

Widespread use of transgenic *Bt* corn hybrids and other genetically altered corn hybrids increases the potential for target insects to develop resistance to the *Bt* endotoxin or to other resistance factors. Consequently, resistance management strategies are necessary to delay the onset of insect resistance and prolong the availability and usefulness of the genetically altered hybrids. Research to determine the most appropriate and effective resistance management strategies is underway. The strategies may vary among different target pests, sources of insect resistance, and geographical region. Insect-resistant corn hybrids likely will become a cornerstone of corn insect management programs.

Other characteristics of corn plants, not necessarily selected as insect-resistance factors, also may influence the degree of injury caused by some insect pests. For example, hybrids with good stalk strength and resistance to stalk rot organisms tolerate injury caused by several stalk-boring insects. Hybrids that have extensive root systems or the ability to compensate for root injury tolerate infestations of corn rootworm larvae. Hybrids that germinate and grow well when soils are cool are less likely to be injured by seed and seedling pests (seedcorn beetles, seedcorn maggots, white grubs, billbugs, wireworms) because the plants are susceptible for a shorter time. Also, some corn hybrids compensate for plants lost to seedling insects by producing sufficient yield from the plants that remain.

Sampling and Monitoring

Surveying insect population levels is essential for the success of an insect management program. Surveys can be conducted within fields (scouting) or over a large area (monitoring). All IPM decisions should be based upon accurate and current information regarding pest densities and conditions.

A grower or consultant can use many different sampling techniques to determine the status of insect pests within a field, depending upon the insects present and the condition of the crop. The presence and/or density of insects may be assessed by observing plant injury (e.g., percentage defoliation, reductions in plant population, percentage of plants with injury) or by counting the insects present. Depending upon the pest and its stage of development, counts can be made on intact plants in the field (e.g., European corn borer eggs, adult corn rootworms), or plants may have to be dissected or dug up to count the insects (e.g., European corn borer larvae, corn rootworm larvae). In some instances, traps may be used to determine relative pest densities (e.g., wireworms, corn rootworm adults). The type of trap used depends upon the insect and stage of development to be sampled. For example, wireworm larvae are attracted to germinating seeds in bait traps placed in the soil, and corn rootworm adults are attracted by yellow sticky traps in the field.

Other techniques may be used to determine the presence or abundance of some insects in an area without making counts in specific fields. Although these monitoring techniques do not provide an absolute density (numbers of insects per plant or per acre), they reflect population trends and indicate when scouting for a pest should begin. However, they seldom provide enough information to signify when a control measure should be applied. Blacklight traps can be used to monitor several moth species, and moth captures may indicate when scouting for eggs or larvae in fields should begin and end. Pheromone traps also may be used as monitoring tools. Pheromone traps that are baited with a chemical sex attractant are available for several important corn insect pests (black cutworm, corn earworm, European corn borer, southwestern corn borer). Informal monitoring techniques (e.g., observations made when a grower walks through fields) can be useful for recognizing insect activity.

Regardless of the sampling technique used, an understanding of the proper sample size and sampling pattern within a field is important to obtain reliable information. Insects rarely are distributed uniformly within a field; the pattern of distribution varies among insect species. These differences influence how many plants and how many locations within a field must be examined to obtain a reliable estimate of the level of pest infestation. Available information for each pest is discussed in the appropriate sections.

Role of Weather in Monitoring for Insect Pests and Predicting Their Activity

Temperature and Insect Growth. Insects are cold-blooded animals; their growth and development are influenced by temperature. Within certain limits, the growth of insects is related directly to temperature, so insect development usually can be predicted by keeping track of temperature accumulations, a more reliable method than predicting insect development based upon the calendar.

Optimum development of insects usually occurs within a range of temperatures. Most insects have a minimum developmental temperature and a maximum developmental temperature, below and above which, respectively, insect development ceases. Heat units (or degree-days) are a measure of the amount of temperature available for insect growth. Heat units may be calculated simply as the average daily temperature minus the minimum developmental temperature, or the base temperature ([maximum temperature + minimum temperature] / 2 – base temperature). For example, the base temperature for development of European corn borers is 50°F (10°C). If the daily maximum temperature was 80°F (27°C) and the daily minimum temperature was 60°F (15.7°C), a total of 20 FDD (Fahrenheit degree-days; 11.4 CDD [Centigrade degree-days]) occurred on that day ([80 + 60 / 2] – 50 = 20 FDD; [27 + 15.7 / 2] – 10 = 11.4 CDD).

This simple equation works well when both the minimum and maximum temperatures are above the base temperature. However, temperatures may fluctuate during a 24-hour period, so other more complex equations and computer programs may be needed to predict insect growth and development accurately.

The minimum and maximum developmental temperatures and the number of FDD required to complete development of different stages of growth are known for only a few insect pests that attack corn, e.g., black cutworm, corn rootworms, European corn borer, southwestern corn borer, and a few others. This information often is used to predict the occurrence of damaging stages and to time scouting activities. Computer programs that accept temperature input and insect information are used to predict the development of European corn borers throughout the United States.

Day Length. Day length, or the ratio of light and dark, influences the development of many insects. Day length may influence whether an adult insect reproduces or prepares for diapause, or whether an immature insect molts to the next stage or enters diapause. Temperature and food quality also may influence the timing of diapause. For example, European corn borer larvae enter diapause based upon the interaction between temperature and day length.

Influence of Weather on Insect Movement. Several important corn insect pests do not overwinter in the area where they cause damage during the growing season. In the spring, these insects move north from their overwintering sites and establish populations. Several groups of insects, including some moths, aphids, and leafhoppers, migrate northward annually.

Spring weather patterns influence whether certain areas are colonized by migrating insects. For example, the black cutworm overwinters in the Gulf Coast region of the United States and farther south. Adults fly northward in the spring, aided by prevailing winds and storm fronts, and females lay eggs on winter annual weeds and crop residue in midwestern fields destined for corn production. Depending upon the timing and direction of weather fronts in the spring, flights are more or less abundant in different parts of the Midwest. A network of pheromone traps distributed throughout the Corn Belt each year is used to predict the occurrence of cutworm activity in different regions. Based upon degree-day accumulations after a critical number of adults is captured in the traps, predictions about timing of scouting for black cutworm injury can be made.

For other migratory insects, informal monitoring may be useful as an early alert for potential problems. Outbreaks of migratory pests to the south often are the first warning that an insect may soon appear in an area.

Decision Making

After sampling for an insect pest and estimating pest density or level of injury within a field, one must decide whether a control action is justified. The concept of an economic injury level (EIL) was developed to provide objective guidelines for making informed decisions about insect control. An EIL is a pest density or level of injury that will result in a yield loss equal to the cost of controlling the pest. It is based upon several factors, including the market value of the crop, control costs, and the yield loss caused per insect. One equation for calculating an EIL is:

$$\text{EIL} = (\text{control costs [\$]}) \div (\text{crop market value [\$]} \times \text{yield loss per pest} \times \text{\% control}).$$

The EIL changes if any of these factors change. For example, if the market value of a crop increases, the EIL decreases. In other words, a grower would be inclined to control an insect pest at a lower density if the market value of the crop increased. By their very nature, EILs are adaptable to changing economics of crop production and pest control.

In recent years, environmental costs have been added to some of the equations for calculating economic injury levels. As an added cost to a grower, environmental disruption caused by an insecticide application might increase the EIL. Although environmental costs are difficult to assess, growers throughout the Corn Belt have offered opinions about environmental costs associated with different hazards posed by insecticides (see Higley and Wintersteen 1992, reference number 119). These costs also vary among insecticides, depending upon the characteristics and use pattern of an insecticide.

The economic threshold is the pest density or level of injury at which a control tactic must be implemented to prevent an increasing pest population from reaching the EIL. Economic thresholds often are established arbitrarily at 80% of an EIL to allow for time to implement control procedures.

Development of EILs requires considerable research to establish the relationship between insect density or level of injury and yield loss. For some insects (e.g., European corn borer), well researched EILs are available. However, EILs have not been determined for many corn insect pests. When EILs are lacking, guidelines (or nominal thresholds) based upon field experience or limited research may be available. Nominal thresholds are used frequently by crop consultants and are available in many university Extension publications. Nominal thresholds may be valuable for avoiding unnecessary insecticide applications when insect pest densities are low. Although appropriate research data may not be available, experience usually allows for some adjustment of nominal thresholds when the cost of control or market value of the crop changes.

The EILs and economic thresholds focus primarily upon the economic aspects of pest management decisions because they are the easiest to quantify. Environmental effects of control efforts (as indicated previously), availability of equipment and trained personnel to apply pesticides or other control measures, and the economic, environmental, and social goals of the producer also are important considerations.

Developing an IPM Program

The information in this handbook and from other sources (e.g., state Extension publications) can be used to develop an insect management program for the most common insect pests in a region, on a farm, or in a field. The following steps are important:

- identify primary and sporadic pests,
- understand pest biology and behavior,
- understand crop management options,
- select pest management options, and
- evaluate the effectiveness of the program, and modify if necessary.

Identify Primary and Sporadic Pests. Some pests occur in damaging numbers almost every year, others occur more sporadically, or are always present but rarely at damaging levels. Economically important primary pests should be identified, and an insect management program that incorporates appropriate nonchemical, preventive tactics should be developed. Plans to monitor for primary pests and respond with remedial tactics, if warranted, also should be developed. Pests that injure corn infrequently may not warrant specific preventive practices. However, appropriate sampling procedures, economic thresholds, and remedial tactics should be identified and used if needed.

For example, throughout much of the Corn Belt, the key insect pests of corn are the corn rootworm complex and European corn borer. In some regions, black cutworm is considered to be a key pest, but its sporadic occurrence designates it more appropriately as an occasional pest. An insect management program for these pests might be designed to prevent corn rootworm larval damage by rotating corn and soybeans annually, or by applying a soil insecticide to corn planted after corn if the preceding year's density of rootworm adults exceeded the established economic threshold. In addition, a grower may choose to plant a corn hybrid that is resistant to European corn borer, possibly a transgenic *Bt* corn hybrid. Early planting helps establish the corn crop and reduces the potential for injury caused by black cutworm. A scouting program should focus upon the times during the growing season when any of these insect pests are present. A regular scouting program designed specifically to survey the field for black cutworm, first- and second-generation European corn borer, and corn rootworm adults enables a grower to detect the presence of less common pests such as corn flea beetle, stalk borer, and corn leaf aphid. Insecticides would be applied only if the density of the pest exceeded established economic thresholds and hazards posed by the insecticide were minimal.

Understand Pest Biology and Behavior. Growers should be able to identify the major and minor pests and their important natural enemies and understand their behaviors and life histories. Information about identification, pest status, and life history is provided in the sections of this handbook related to specific pests. Beneficial organisms and the insect pests with which they are associated also are discussed elsewhere in this handbook. Useful information can be obtained locally from Extension (bulletins, newsletters, workshops, etc.) and from other sources (seed dealers, pesticide suppliers).

Understand Crop Management Options. Both preventive and remedial control tactics are identified for most of the insects discussed in this handbook. The effects of crop management practices on the potential for insect infestations are explained. For example, planting corn early reduces the potential for infestation by black cutworm but increases the potential for infestation by first-generation European corn borer. When known, the effects of crop rotation, tillage, and other crop management activities on insect populations are discussed. The availability and usefulness of resistant corn hybrids also are addressed. Seed corn dealers can provide additional information about levels of insect resistance in locally adapted corn hybrids.

Depending upon the specific crop management practices suited for a region and cropping system, adoption of some of these practices may not be easy or reasonable (e.g., limitations related to crop rotations, lack of a resistant hybrid adapted to a particular region). Consequently, a grower may wish to seek additional information about the impact of cropping practices on insects found in corn.

Select Pest Management Options. After the primary and sporadic pests in a region have been identified and studied and management options have been evaluated, a grower should develop a pest management plan that focuses primarily upon nonchemical preventive tactics. Insecticides should be used only to supplement a completely integrated pest management program that includes the use of insect-resistant corn hybrids and cultural and biological control methods. Pest management options that are effective, economical, and compatible with specific cropping systems should be implemented. A grower also should determine whether he or she has the time and expertise to scout cornfields for pest problems, or whether a crop consultant should be hired.

Economic thresholds for some insects that attack corn are included in the appropriate sections of this handbook. However, economic thresholds may vary from region to region, so locally available information should be consulted. Application of an insecticide is an option available for most insects that attack corn. However, not all insecticides are labeled for all insects, so knowledge about the registration and effectiveness of different insecticides is important.

Evaluate the Effectiveness of the Program, and Modify if Necessary. After a pest management plan has been developed and implemented, its effectiveness should be evaluated, and the plan should be modified if necessary, to accommodate changes in pest species or pest control technology. Pest species may change in importance over time for many reasons or they may develop resistance to chemical, cultural, or biological control practices. Also, new pest control technologies may become available, or existing technologies may cease to be available or effective.

Corn IPM in the Future

Several aspects of corn insect management are changing rapidly and will continue to change in the future. Transgenic corn for insect control was available commercially for the first time in 1996 when *Bt* corn hybrids for European corn borer control were introduced. In the near future, transgenic or other genetically altered hybrids with resistance to corn rootworms and other insects likely will become available. New classes of insecticides with novel modes of action and less potential to disrupt the environment are being developed. Bait formulations that contain very small amounts of insecticide are available for control of adult corn rootworms. Applications of site-specific management (yield mapping, variable rate technology, remote sensing) are being implemented in corn production, and their applications to corn insect management are being studied.

Insect pests of corn also are changing, and these pests may adapt to management practices in time. In Nebraska, localized populations of western corn rootworms have developed resistance to organophosphate and carbamate insecticides, resulting in reduced levels of control in some areas. In some areas of Illinois, Indiana, Michigan, and Ohio, females from populations of western corn rootworms are laying eggs in soybeans, resulting in economic damage to first-year corn planted after soybeans.

The need to stay current and knowledgeable about new insect management tactics and research results and to incorporate this information into corn production systems that are profitable and environmentally sound has never been more important.

Selected References. 119, 172, 186, 235

Corn Pests

Insects and Related Arthropods

By B. R. Wiseman and Randall A. Higgins

Several different types of insects and related arthropods attack corn from the time the seed is planted through harvest. Some pests attack more than one plant part or cause injury at more than one crop growth stage. Knowledge of corn plant physiology and development is helpful for predicting the severity of injury and may suggest ways to alleviate the potential for reduced yield. Corn, however, is less able than soybeans to compensate for mid-to-late season defoliation and injury to reproductive tissues. Plant vulnerability to insect injury also depends upon geographic location and climatic factors. For instance, in the southern United States, corn is attacked by more pests and in greater numbers than corn grown in more northern Corn Belt regions.

Generally, insects and their relatives cause four types of injury to corn plants: (1) injury to seeds and seedlings; (2) injury to roots; (3) injury to leaves, tassels, and stalks; and (4) injury to ears. Injury caused by specific pests is discussed in the sections on individual pests. Refer to the diagnostic key on pages 25–29 for help with recognizing injury caused by insect pests.

Injury to Seeds and Seedlings. Cold soil slows germination and plant emergence, extending the time available for subterranean pests (wireworms, white grubs, seedcorn maggot, seedcorn beetles) to feed upon and injure the plants. Injury to seeds often results in hollowed out seed coats as the only evidence that the seeds were planted. Plant death early in the season creates gaps in the stand, the severity of which depends upon the proximity and density of surviving plants. Symptoms of injury to corn seedlings include rolled leaves; discolored, sometimes purple leaves; wilted plants, possibly with dead whorls; stunted or dead plants; poor stands; and tillering of young plants. These injury symptoms are caused by several pest species, including wireworms, white grubs, chinch bug, corn root aphid, stink bugs, billbugs, cutworms, lesser cornstalk borer, sod webworms, hop vine borer, and stalk borer. Early season injury to foliage, such as that caused by flea beetles, may not be economically damaging if environmental conditions are favorable for corn growth and development. Early season tolerance to insect injury lasts until emergence of the fifth or sixth leaf because the growing point of a corn plant remains below ground and is relatively protected from serious injury.

Injury to Roots. Injury to corn roots can be caused by white grubs, corn rootworm larvae, and several other pests. Rootworm larvae can damage the main root system severely. Root tips may appear brown and chewed back (pruned), and larvae may tunnel into larger roots, resulting in a physiological reduction in yield because the plants cannot take up moisture and nutrients easily. Also, brace roots may be severely damaged, resulting in lodged plants with gooseneck-shaped stalks. Yield losses occur because photosynthetic efficiency declines as optimal leaf display is altered by lodging.

Injury to Leaves, Tassels, and Stalks. Insects may injure corn leaves from the seedling stage through maturity. Injury to seedlings caused by billbugs appears as irregular holes in unfolded leaves or as a series of transverse, oblong holes in expanded leaves. On larger plants, pinholes, small and large lesions, ragged leaves, and shredding of leaves can be caused by European corn borer, fall armyworm, southwestern corn borer, stalk borer, and corn earworm larvae. Similar leaf injury is caused by grasshoppers.

European and southwestern corn borers feed on several corn plant parts, depending upon the time of attack. Although young larvae typically cause minor injury to leaves, older larvae tunnel within stalks, damaging the plant's vascular system. Injured plants transport water and carbohydrates less efficiently, so the plants may be smaller and have fewer leaves and weakened stalks. Injured plants also are more susceptible to stress, particularly if water is not readily available. After feeding on leaves in the whorl, stalk borers and southwestern corn borers feed in the nodes and tunnel in stalks, resulting in plant lodging and/or ear drop. Southwestern corn borers also girdle the stalks. Broken tassels can be caused by tunneling of corn earworm, fall armyworm, and European corn borer larvae. If injury to the stalk or shank occurs during reproductive growth stages, injured plants may not be able to reallocate photosynthates stored within the stalk to kernels, further reducing yields. Weakened plants are less able to tolerate other stresses such as stalk rots and spider mites.

During the reproductive growth stages of field corn, European corn borers may tunnel within stalks, ear shanks, and cobs, and may feed on the kernels. Yield losses caused by second-generation corn borers are correlated with time of attack relative to crop growth stage. Yield losses caused by their attack during late vegetative growth stages and early reproductive development may be 5–6% per borer per plant. Infestations that occur after the kernel reaches the dough stage may result in 2% or less loss per borer per plant, as long as ears remain attached to the stalks.

The impact of insect-induced injury to corn foliage depends upon the health and development of the entire stand. Individual plants might suffer little from complete foliage loss (for example, cutting by cutworms) early in the season if these plants are grown in isolation. However, when injured plants are surrounded by more vigorous, healthy plants, they may not be able to contribute proportionally to yield. When several adjacent plants are defoliated early in the season, weeds may take advantage of abundant uncaptured sunlight and further stress the crop.

Injury to Ears. Insects that attack corn ears may eat the silks, injure the kernels, or tunnel in the shanks, resulting in dropped ears. The corn earworm, fall armyworm, and European corn borer feed on developing kernels, causing yield loss. Japanese beetles and grasshoppers feed on emerging silks and kernels near the ear tip. Adult corn rootworms also may clip silks, potentially hindering pollination, a particular problem in the production of seed corn. However, the importance of this problem to vigorous hybrids may be overstated because silk elongation frequently offsets the loss of silk caused by insect feeding.

Injury to the husk or ear caused by corn earworms, corn borers, and other large insects creates wounds through which secondary pests (arthropods and pathogens) penetrate and feed on or infect newly formed kernels. Ear rots and several different insects, such as "picnic beetles" attracted by fermenting grain, add to the losses caused by the caterpillars. Because corn plants have limited ability to compensate for injury by increasing seed size, insect injury to ears usually results in yield loss.

Insect pests can reduce the potential number of rows of kernels (usually set by stage V12) and kernels per row (set just before silking), thereby limiting the number of kernels produced per plant. Stress that occurs during and after the period when maximum potentials are determined can result in fewer rows of kernels, kernels per row, and seed size. The 4-week period from 2 weeks before through 2 weeks after silking is the period of growth with the greatest vulnerability to yield loss.

Recent data have shown that the physiological effects of one European corn borer tunneling within the ear shank may be more serious, in terms of yield losses, than losses caused by a single borer tunneling within the corn stalk. Physiological losses are compounded by losses at harvest if shank tunneling causes the ears to fall from the plants. Fifty to 80% of shank-infested ears may drop from plants when harvest is delayed. Shank-infested ears generally do not begin falling from corn plants until about 3 weeks after black layer formation. Thus, early harvest of shank-infested fields seems warranted.

Selected References. 26, 40, 68, 97, 190, 213, 220

Weeds

By Michael J. Horak and Loren J. Moshier

Weeds and crops require the same resources and compete for water, nutrients, light, and other essentials of plant growth. Weeds are considered undesirable because they reduce crop yield and quality, interfere with harvest operations, and may harbor insect pests or serve as alternate hosts for insects and pathogens. Weeds also increase the time and costs of crop production.

The cost of weeds in crops derives from the loss of crop yield due to weed competition and the cost of controlling weeds. Losses in corn yield caused by weed infestation in the United States from 1989 to 1991 were estimated to be $1 billion per year when herbicides were used with best management practices. With only best management practices and no herbicides, the loss was estimated to be $19.6 billion per year. Corn growers spend approximately $1 billion per year for herbicides and at least an additional $1 billion per year in labor, fuel, and machinery costs for chemical and mechanical weed control. Based upon nearly 78 million acres (30 million hectares) of corn production in the United States, total costs associated with weeds are almost $40 per acre.

Weed species that infest corn vary by region in frequency and severity. It is not uncommon for more than 20 species to occur in the same field. However, usually just a few species are most problematic in a given year. Some of the common weed species in corn are panicums (*Panicum* spp.), pigweeds (*Amaranthus* spp.), Johnsongrass (*Sorghum halepense*), shattercane (*Sorghum bicolor*), foxtails (*Setaria* spp.), velvetleaf (*Abutilon theophrasti*), common lambsquarters (*Chenopodium album*), sedges (*Cyperus* spp.), common cocklebur (*Xanthium strumarium*), morningglories (*Ipomoea* spp.), and crabgrasses (*Digitaria* spp.).

Most corn in the United States is grown in a monoculture with an attempt to remove weeds that germinate at the same time as the crop and for several weeks after crop emergence. The monoculture system provides a uniform environment with concentrated resources that may encourage insect infestation. The predisposition of monocultures to insect outbreaks is well known. One theory suggests that outbreaks of insect pests may occur because populations of beneficial insects are reduced by the lack of alternative food sources, shelter, and breeding sites in the monoculture environment. On the contrary, fields with a diversity of weed species may have greater numbers of predaceous arthropods. Consequently, outbreaks of some insect species would occur more frequently in weed-free fields than in fields with some weed diversity.

Weed species may influence crop–insect interactions. Weeds may serve as a site of refuge or as alternate hosts for pest species, species that transmit disease pathogens, and beneficial species. Weeds may enhance or inhibit the attractiveness of the crop and may alter the physical environment around crops, possibly affecting pest infestation (either by enhancement or inhibition). In addition, some insect pests may reduce the competitiveness of the crop and, thus, encourage weed growth and interference.

Although much of the evidence that weeds affect other pest problems is indirect or observational, some weed–insect interactions in corn have been documented. In Massachusetts, infestation of European corn borers in sweet corn ears increased from 20% in weed-free plots to 40% in weedy plots. Broadleaf weeds (lambsquarters and pigweed) positively influenced infestation more than grasses (panicum and crabgrass). In Iowa and Louisiana, the time of weed removal significantly affected (both positively and negatively) seedling damage caused by black cutworm larvae, depending upon the developmental stage of the insect. The researchers affirmed the importance of proper weed control combined with monitoring insect development to limit corn injury. In Indiana, heavy foxtail infestations reduced densities of corn rootworm larvae and adults. However, foxtail competition can reduce corn yield. In Florida, occurrence of fall armyworms was greater in weed-free habitats than in corn infested with weeds. In the same study, the presence of weeds had no effect on the level of injury caused by corn earworms.

Weed–insect–crop interactions are poorly understood. Any beneficial effect of leaving weeds uncontrolled to help manage an insect pest must be considered in light of the detrimental effects of the weeds on the crop.

Selected References. 6, 7, 13, 30, 135, 254

Corn Diseases and Their Relationship to Insect Management

By G. P. Munkvold

Diseases are a chronic source of yield loss and quality reduction wherever corn is grown. Diseases of corn can be categorized as foliar diseases, root and stalk diseases (including seedling blights), and ear and kernel diseases (including storage molds). Some pathogens cause more than one type of disease (e.g., stalk rot and ear rot). Fungi comprise the most important group of corn pathogens. Important corn diseases in the United States include stalk rot caused by *Fusarium* species or *Colletotrichum graminicola* (anthracnose); seedling blight caused by *Pythium* or *Fusarium* species; gray leaf spot (*Cercospora zeae-maydis*); northern corn leaf blight (*Exserohilum turcicum*); common maize rust (*Puccinia sorghi*); eyespot (*Aureobasidium zeae*); several virus diseases; nematodes (several genera); and ear or kernel rots caused by *Diplodia maydis* or species of *Fusarium*, *Aspergillus*, or *Penicillium*.

Insects often play a major role in the occurrence of many corn diseases. Insects affect diseases primarily by acting as vectors of plant pathogens. All types of pathogens (fungi, bacteria, nematodes, viruses, mycoplasmas) can have insect vectors, although vectors are important for only a few of the major corn diseases. Insects also influence the occurrence of diseases by wounding plant tissue, providing a means of entry for plant pathogens that cause several different diseases in corn. Finally, when plants are stressed by insect injury, they are more vulnerable to infection and development of disease.

Foliar diseases caused by fungi are not affected much by insect activity. Most fungal pathogens produce abundant spores that usually are windblown or splashed onto leaf surfaces. Although insects carry the spores of some of these pathogens, insects are not a major dispersal mechanism for the spores.

Stewart's disease, an important foliar disease caused by the bacterium *Erwinia stewartii*, is transmitted by insects. The major mechanism for dispersal of the bacterium is the corn flea beetle. The bacterium overwinters in the gut of the beetle and infects corn plants as the beetle feeds on leaves. Stewart's disease has a wilt phase early in the season that may be fatal to susceptible genotypes and a more common leaf blight phase later in the season. Planting resistant genotypes is the only method of managing the disease directly. However, controlling the flea beetle consistently reduces the incidence of the disease. Application of a foliar insecticide when the density of flea beetles exceeds established thresholds is an economically viable method of control, especially in corn inbreds that are very susceptible to Stewart's disease. Some systemic soil insecticides or seed treatments also reduce the incidence of Stewart's disease by inhibiting feeding by corn flea beetles.

Insects also play a major role in the spread of viral diseases of corn, such as maize dwarf mosaic virus (MDMV). More than 20 different species of aphids, including the corn leaf aphid, the greenbug, and the green peach aphid can act as vectors for this virus. Aphids typically acquire MDMV from Johnsongrass, where the virus overwinters, and spread the virus to corn plants. Other viral diseases of corn are spread by beetles, leafhoppers, planthoppers, and the wheat curl mite. A recently discovered viruslike disease called High Plains disease also may be spread by the wheat curl mite. Although the causal agent for this disease probably is a virus, its identity has not been verified. Viral diseases of corn usually are managed by cultural practices and by the use of resistant corn hybrids. Although insecticides for control of the vectors have been recommended, the economic benefits are questionable.

Insects influence several root and stalk diseases of corn. Although pathogens can infect the roots of seedlings that have not been injured by insects, wounds caused by insects such as the seedcorn maggot often become infected by soilborne pathogens. However, the insect is not involved in spreading the pathogen.

Root and stalk rots of older plants also are affected by insect activity. Injury caused by European corn borer and corn rootworm larvae increases the incidence of all types of stalk rot. Wounds caused by the insects' feeding provide a means of entry for the fungi, and injury to the roots and stalks causes stress that predisposes the plants to the development of stalk rot. Additionally, European corn borer larvae carry spores of *Fusarium* species from the plant surface to the interior of stalks where stalk rot is initiated. Management of corn borers and rootworms can play a major role in management of corn root and stalk rots.

Corn ear and kernel rots are affected by insect activity in a manner similar to the stalk rots. Some of the same fungi that cause stalk rots, such as the *Fusarium* species, also infect corn kernels. Symptoms of *Fusarium* ear rot, caused by *F. moniliforme*, *F. subglutinans*, and other species, are correlated with the amount of injury to the ear caused by European corn borer and corn earworm larvae. Insect activity may be the most important factor affecting the occurrence of these diseases. European corn borer larvae carry *Fusarium* spores from the plant surface to the kernels and create entry wounds for the fungi. The result is an increase in disease symptoms and in symptomless infection of the kernels. Because these fungi can produce mycotoxins, management of both symptomatic and symptomless infection is important. Ear and kernel pathogens such as *Aspergillus*, the producer of aflatoxin, also are associated with insect injury to ears. Thrips and sap beetles have been implicated as vectors and/or wounding agents for the development of ear and kernel rots of corn.

The European corn borer probably is the most important insect associated with diseases of corn. Because corn borers have such a strong influence on stalk and ear rots, management of corn borers provides detectable benefits in yield and quality by reducing the incidence of rots. Reductions in the amount of damage caused by European corn borers, either by application of insecticides or planting borer-resistant hybrids, significantly reduces the incidence of both stalk rots and ear rots.

The influence of insects on diseases continues after corn grain is stored. Insect pests such as grain weevils that feed on the kernels cause an increase in the incidence of storage molds caused by fungi primarily in the genera *Aspergillus*, *Penicillium*, *Cladosporium*, and *Fusarium*. Insect activity also may affect temperature and moisture conditions in stored corn, promoting proliferation of storage molds. Insects in stored corn carry spores of the storage molds, but their importance as a source of inoculum is not clear because many corn kernels are contaminated or infected before they are stored. Nevertheless, insect management in storage reduces the potential for the occurrence of storage molds.

Management of corn diseases can be accomplished by integrating several tactics, including genetic resistance, crop rotation, and other cultural practices. Because of the complex interactions among disease pathogens, insects, and the host plants, and the particularly strong influence of corn insects on diseases, insect management is an important component of integrated management of corn diseases.

Selected References. 142, 224, 226

Birds and Mammals

By Richard A. Dolbeer

Several species of birds and mammals cause substantial losses to sprouting and ripening corn. Some birds, especially crows (*Corvus brachyrhynchos*) and grackles (*Quiscalus quiscula*), uproot sprouting corn to feed on germinating kernels. Red-winged blackbirds (*Agelaius phoeniceus*) and grackles destroy more than 360,000 metric tons of ripening field corn annually in the United States and Canada. Losses can be particularly severe in sweet corn because even slight bird damage renders ears unmarketable as fresh produce. Most loss of ripening corn to birds occurs within 5 mi (8 km) of marshes where blackbirds roost in late summer. These roosts sometimes contain more than 1 million birds. Blackbirds typically slit open husks with their bills to puncture kernels in the milk stage, leaving the kernel coats attached to the cob. Subsequent secondary damage caused by mold often occurs. Blackbirds frequent fields heavily infested with insects such as rootworms and corn earworm, feeding extensively on these insects as well as on corn. Suppression of insect populations in corn makes these sites less attractive to blackbirds. The use of corn varieties with long, thick husks also reduces the extent of injury. Blackbirds can be frightened from fields with propane exploders and shotguns.

Rodents such as thirteen-lined ground squirrels (*Spermophilus tridecemlineatus*), deer mice (*Peromysus maniculatus),* house mice (*Mus domesticus*), and prairie and meadow voles (*Microtus* spp.) will feed on germinating corn seeds. Losses can be substantial, requiring replanting, especially in no-till plantings and in cropping systems that include cover crops. Vegetation provides ample food to encourage reproduction and overhead cover for safety from predators. The lack of tillage allows undisturbed establishment of burrows. Damage occurs in these fields after preemergence herbicide applications destroy much of the food resources of the rodents, forcing them to feed upon corn seeds and young plants. Woodchucks (*Marmota monax*) also feed on sprouting corn plants along field edges. Alternative feeding, chemical repellents, habitat modification, and poison baits (registered in some states) are recommended for management of rodents in corn.

Injury to ripening ears by populations of white-tailed deer (*Odocoileus virginianus*) and raccoons (*Procyon lotor*) is an increasing problem. Deer typically nip off ear tips, whereas raccoons chew through husks, often pulling plants down. Management of deer and raccoon populations through hunting or trapping is the best approach for reducing the level of injury.

Selected References. 70, 71, 72, 73

Identification of Insects and Diagnosis of Injury

Kevin L. Steffey, Marlin E. Rice,
and Michael E. Gray

Effective management of corn insect pests begins with accurate identification of the pests and reliable diagnosis of pest injury. Identification and diagnosis can be based upon three types of evidence: (1) morphology—an insect's form and structure; (2) indications of an insect's presence—e.g., symptoms of injury; and (3) situation—the time of occurrence or location of a pest or its injury in a field or on a plant. A combination of the three types of evidence offers the greatest potential for accurate identification and diagnosis. Descriptions and color photographs of insects and the injury they cause and details about their time of occurrence are provided in the sections pertaining to individual insects in this handbook. However, the list of pests by injury type, the illustrations, and the keys in this section offer a quick method of identifying insects and diagnosing injury in corn.

If the type of injury found in a cornfield can be categorized (e.g., feeding on leaves or tunneling in stalk), the list of corn insects by injury type might provide a first clue about which insect(s) are responsible. The page number after an insect's common name indicates the section in which the insect is discussed in detail.

The keys for diagnosis of injury and insect identification were designed to be used in a stepwise fashion. At each step, two or three choices are presented, one of which should apply to the injury or insect to be identified. The choice made leads to another set of two or three choices, and so on until the name of the insect appears with an illustration. If the illustration does not match the injury or insect, the user should retrace his or her steps to determine where the wrong choice was made. After the injury or insect has been identified with the keys, the color photographs can be studied to verify the diagnosis or identification.

The key for diagnosis of insect injury to corn is presented as a series of choices. Different pests occur at different times of the season, so initial separators are times of the growing season. Other primary separators are the parts of the plants injured. When several insects cause the same type of injury at the same stage of corn development, all of them are listed.

A positive association between the injury to the corn plant and its cause is essential for making accurate pest management decisions. In some instances, injury alone will not distinguish which pest is causing the injury. Every attempt should be made to collect and identify the pest(s). Many species of insects found in corn do not appear in the keys because they are incidental in occurrence. The keys were designed for identification of the most common and important pest insects, and a few unusual corn pests that often are mistaken for more serious pests.

Preceding the key for the identification of insects are generalized illustrations on which certain body regions and structures are indicated. The illustrations of a grasshopper (Mormon cricket), beetle, aphid, caterpillar, and insect leg provide an orientation for some of the terms used in the key. The Glossary provides definitions of these terms.

The key for the identification of insects initially separates pests at or below the soil surface from pests that feed above ground. Other major characteristics used to separate types of insects include the presence or absence, and types of legs and other structures; body shape and general size; types of wings; and color patterns.

After the key for the identification of pests is a series of illustrations of parasitoids and predators of pests. Because a key to the species of natural enemies is impractical, the illustrations provide a representation of these insects' appearance.

Corn Insect Pests by Injury Type

Feeding on Planted Seeds
- Fire ant (p. 83)
- Seedcorn beetles (p. 97)
- Seedcorn maggot (p. 98)
- Slugs (p. 99)
- Thief ant (p. 45)
- Wireworms (p. 117)

Feeding on Roots
- Carrot beetle larvae (p. 54)
- Cornfield ant (p. 45)
- Corn root aphid (in assoc. with cornfield ant) (p. 47)
- Corn rootworm larvae (p. 61)
- Fire ant (p. 83)
- Grape colaspis larvae (p. 85)
- Sugarcane beetle (p. 109)
- Symphylans (p. 110)
- Whitefringed beetle larvae (p. 114)
- White grubs (p. 115)

Feeding on Seedlings: Leaves and/or Stems
- Aphids, some species (p. 46)
- Armyworms (p. 49)
- Billbugs (p. 53)
- Carrot beetle adults (p. 54)
- Chinch bug (p. 56)
- Corn leafhopper (p. 91)
- Cutworms (p. 68)
- False chinch bug (p. 82)
- Flea beetles (p. 84)
- Hop vine borer (p. 88)
- Leafrollers (p. 93)
- Lesser cornstalk borer (p. 94)
- Potato stem borer (p. 88)
- Slugs (p. 99)
- Southern corn leaf beetle (p. 100)
- Stalk borer (p. 106)
- Stink bugs (p. 108)
- Sugarcane beetle adults (p. 109)
- Thrips (p. 111)
- Webworms (p. 112)
- Wireworms (p. 117)

Feeding on Leaves or Stems: Chlorosis or Similar Leaf Discoloration
- Aphids (p. 46)
- Chinch bug (p. 56)
- Corn delphacid (p. 58)
- False chinch bug (p. 82)
- Leafhoppers (p. 91)
- Spider mites (p. 104)
- Stink bugs (p. 108)
- Thrips (p. 111)
- Twolined spittlebug (p. 111)

Feeding on Leaves: Tissue Removal
- Armyworms (p. 49)
- Cattail caterpillar (p. 55)
- Cereal leaf beetle (p. 56)
- Corn earworm (p. 59)
- Corn rootworm adults (p. 61)
- Cutworms (p. 68)
- European corn borer (p. 78)
- Flea beetles (p. 84)
- Grape colaspis adults (p. 85)
- Grasshoppers (p. 85)
- Japanese beetle (p. 90)
- Leafrollers (p. 93)
- Mexican rice borer (p. 104)
- Mormon cricket (p. 94)
- Neotropical corn borer (p. 104)
- Southern cornstalk borer (p. 103)
- Southwestern corn borer (p. 101)
- Sugarcane borer (p. 103)
- Whitefringed beetle adults (p. 114)
- Woollybear (yellow) (p. 119)

Feeding on Leaves: Mining
- Corn blotch leafminer (p. 58)

Feeding in Developing Ears, Silks
- Corn earworm (p. 59)
- Corn rootworm adults (p. 61)
- Corn silk beetle (p. 68)
- European corn borer (p. 78)
- European earwig (p. 82)

Fall armyworm (p. 51)
Grasshoppers (p. 85)
Japanese beetle (p. 90)
Mexican rice borer (p. 104)
Neotropical corn borer (p. 104)
Sap beetles (p. 96)
Southern cornstalk borer (p. 103)
Southwestern corn borer (p. 101)
Sugarcane borer (p. 103)
Western bean cutworm (p. 113)
Woollybear (yellow) (p. 119)

Tunneling in Stalk
European corn borer (p. 78)
Lesser cornstalk borer (p. 94)
Mexican rice borer (p. 104)
Neotropical corn borer (p. 104)
Southern cornstalk borer (p. 103)
Southwestern corn borer (p. 101)
Sugarcane borer (p. 103)
Stalk borer (p. 106)

Feeding on or in Kernels
Angoumois grain moth (p. 44)
Corn earworm (p. 59)
European corn borer (p. 78)
Pink scavenger caterpillar (p. 95)
Wheat curl mite (p. 113)

Transmission of Plant Disease
Aphids (p. 46)
Corn delphacid (p. 58)
Flea beetles (p. 84)
Leafhoppers (p. 91)
Wheat curl mite (p. 113)

Insect Injury to Corn

Early Season Injury: Planting to Early Whorl

Start here

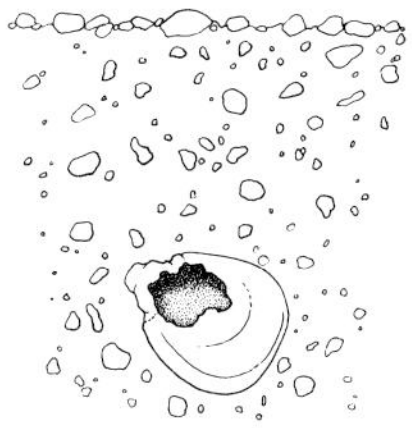

Skips in row where plants failed to emerge; seeds hollowed out or consumed
Red imported fire ant (p.83)
Seedcorn beetles (p.97)
Seedcorn maggot (p.98)
Thief ant (p.45)
Wireworms (p.117)

Germination normal, but seedlings are injured

Seedlings cut off at or just below the soil surface
Cutworms (p.68)
Webworms (p.112)

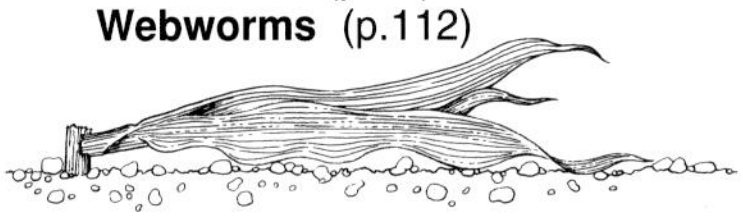

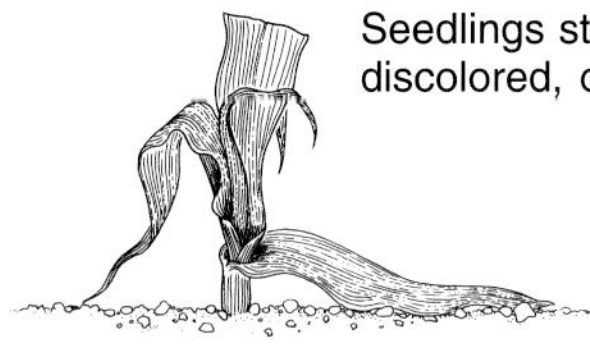

Seedlings stunted, wilted, discolored, or dead

Portions of leaves eaten
(Continued on page 26)

Leaves wilted and/or discolored due to feeding by insects with sucking moutparts

Plants, or parts of plants, yellow or red
Chinch bug (p.56)
Corn root aphid (p.47)
Greenbug (p.48)
Leafhoppers (p.91)

Silver scars on leaves
Thrips (p.111)

Leaves may be discolored, but injury is to stem or roots

Roots injured

Root hairs stripped off, lower stem may be purple, plants stunted
Grape colaspis larva (p.85)
Symphylans (p.110)
White grubs (p.115)

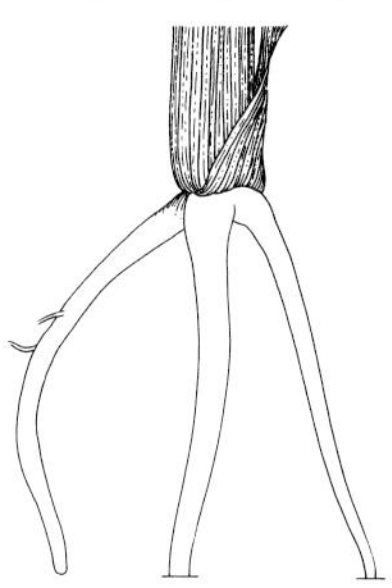

Root tissue eaten, scars or small tunnels evident
Grape colaspis larva (p.85)
Red imported fire ant (p.83)
Southern corn rootworm larva (p.63)
Whitefringed beetle larva (p.114)

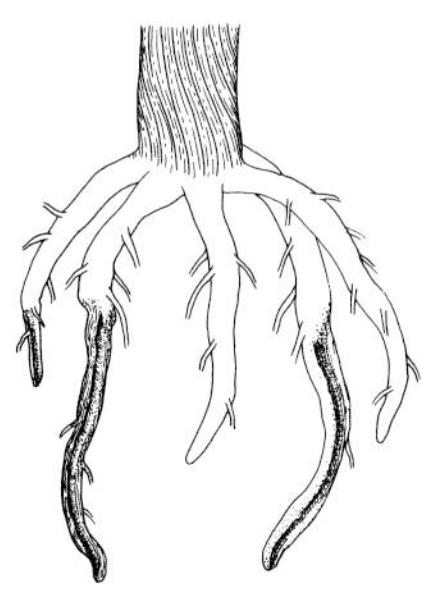

Plants with dead heart, whorl leaf wilted, dying, or dead; may be a hole in base of stem, insect tunneling in lower stem
Carrot beetle (p.54)
Cutworms (p.68)
Hop vine borer (p.88)
Potato stem borer (p.88)
Southwestern corn borer and related stalk borers (p.101)
Stalk borer (p.106)
Sugarcane beetle (p.109)
Webworms (p.112)
Wireworms (p.117)

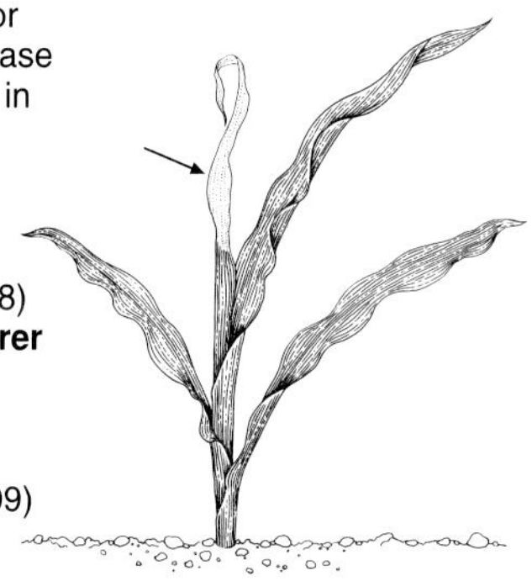

Gougelike hole in side of stalk below ground
Carrot beetle (p.54)
Cutworms (p.68)
Sugarcane beetle (p.109)

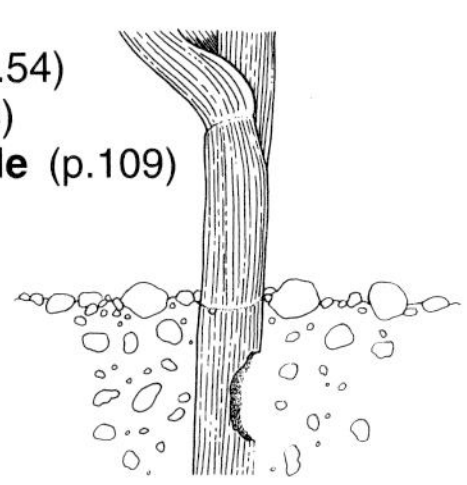

Hole in bottom of young stem, tap root gone
Hop vine borer (p.88)
Potato stem borer (p.88)

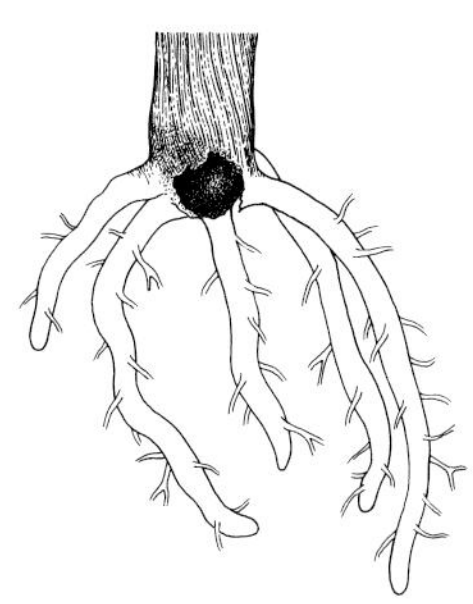

(Continued from page 25)

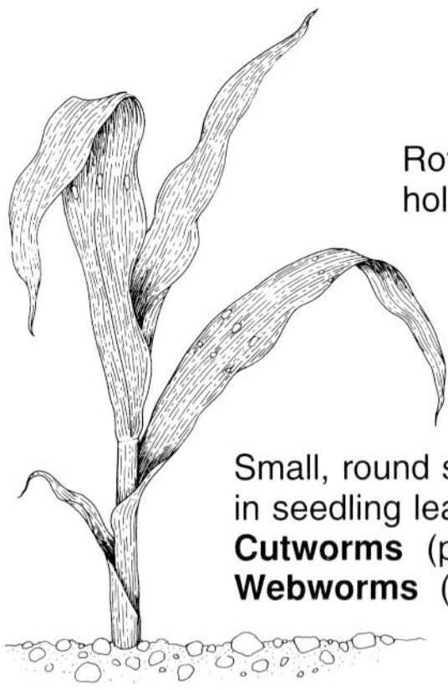

Rows of small, round or oblong holes in leaves of seedlings

Small, round shot holes in seedling leaves
Cutworms (p.68)
Webworms (p.112)

Stunted, malformed plants with rows of oblong holes in leaves; plants tillered at base; leaves often twisted at tips
Billbugs (p.53)

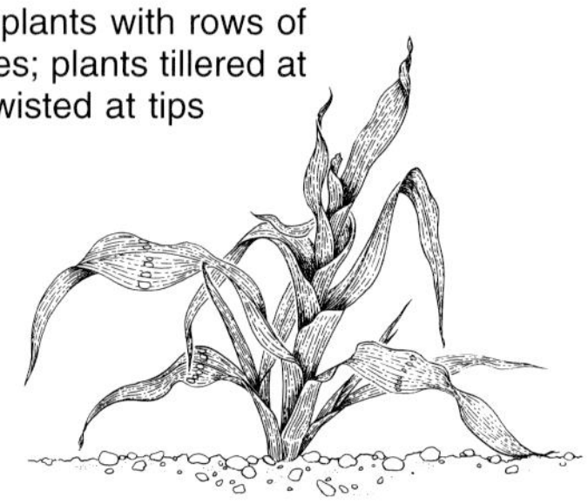

Holes surrounded by necrotic tissue
Stink bugs (p.108)

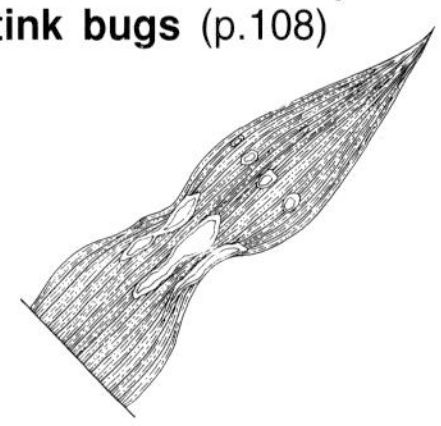

Chunks of leaf tissue eaten or epidermis removed

Leaves not rolled, webbed together

Edges of leaves eaten, leaves rolled, webbed together
Leafrollers (p.93)

Feeding tracks or holes on surface of leaves

Epidermis removed, often in strips, tips of leaves may be wilted
Flea beetles (p.84)

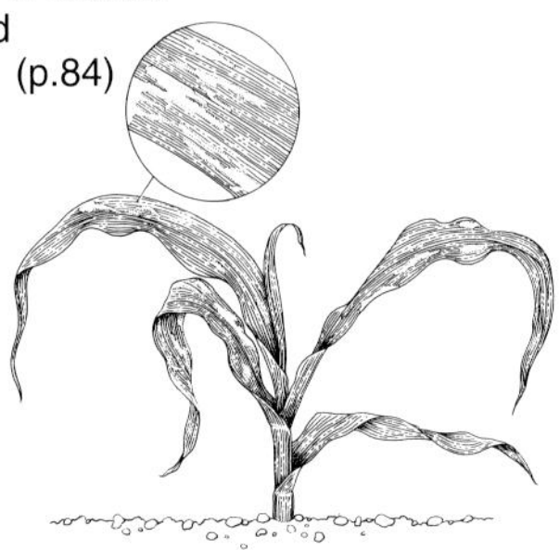

Feeding tracks or holes between veins, bottom epidermis often intact
Cereal leaf beetle (p.56)

Slime trails evident
Slugs (p.99)

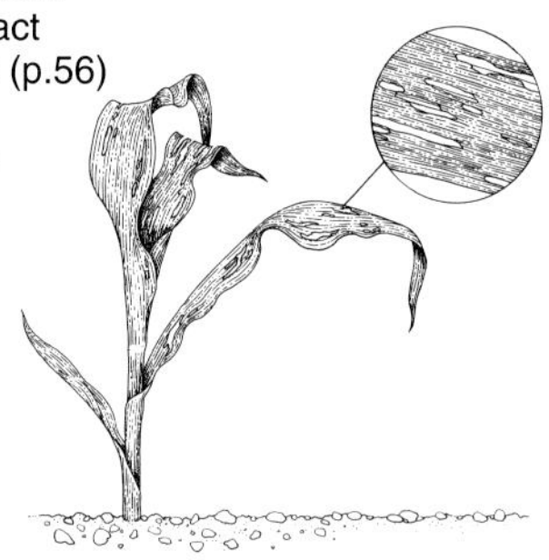

Chunks of leaf tissue eaten

Tops of seedlings cut off squarely
Cutworms (p.68)

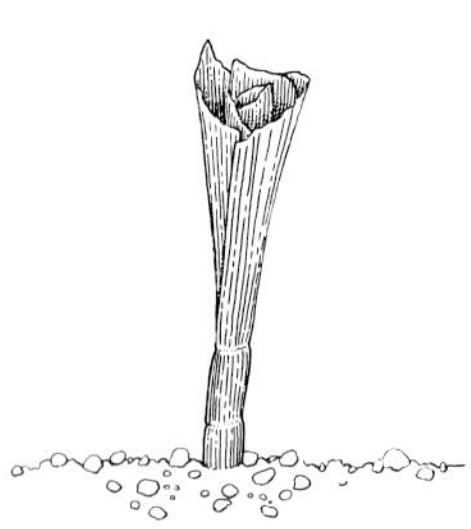

Irregular chunks eaten from edges of seedling leaves
Armyworm (p.50)
Cutworms (p.68)
Webworms (p.112)
Yellowstriped armyworm (p.52)

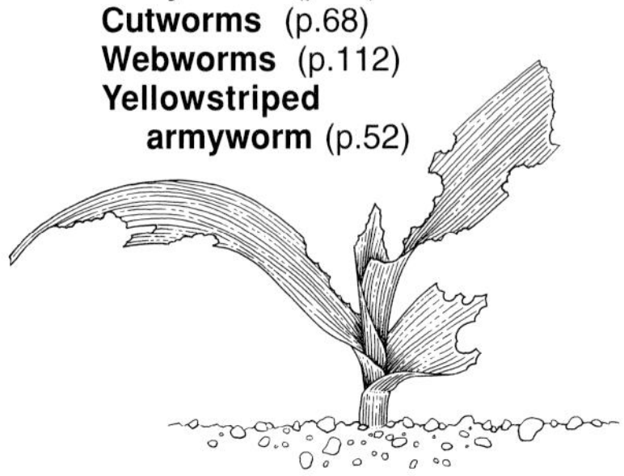

Small notches chewed from seedling leaves and stems
Southern corn leaf beetle (p.100)

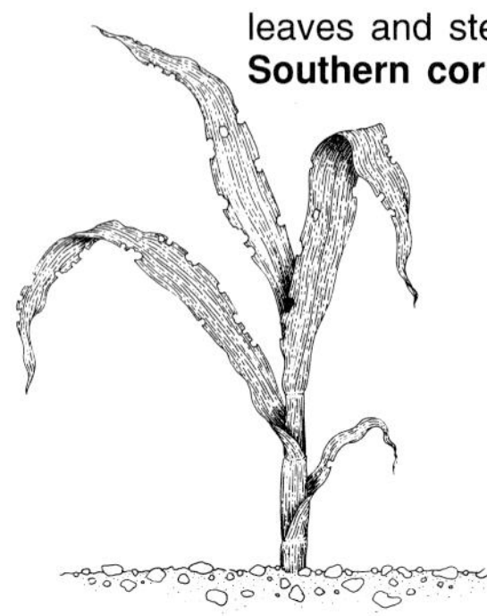

Insect Injury to Corn

Mid- to Late-Season Injury to Leaves, Stalks, Tassels, or Roots

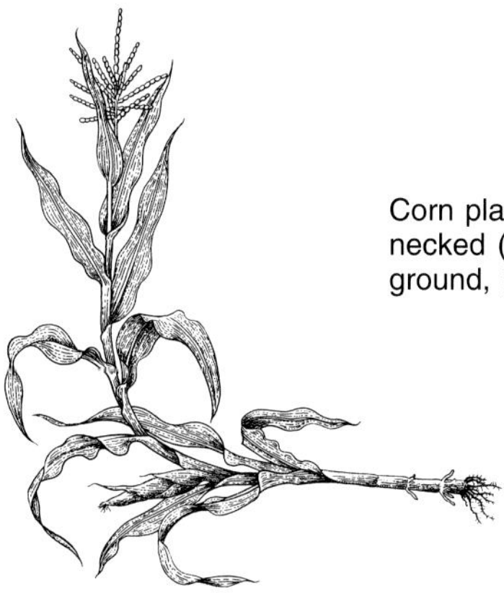

Start here

Corn plants lodged or goose-necked (base of stalk on ground, plant curved upward)

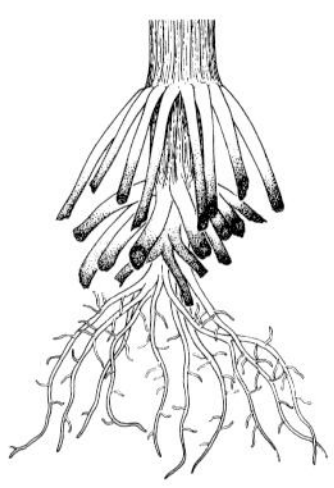

Root system (left) or brace roots (right) severely pruned
Corn rootworm larvae (p.61)

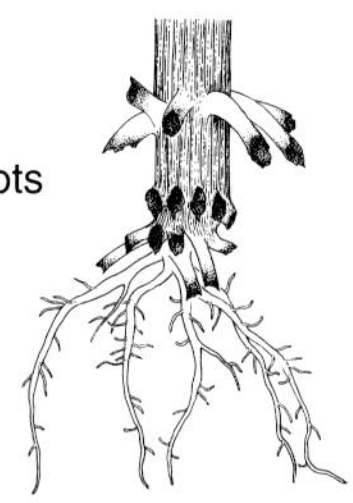

Corn plants not goosenecked, although they may be broken

Injury to stalks

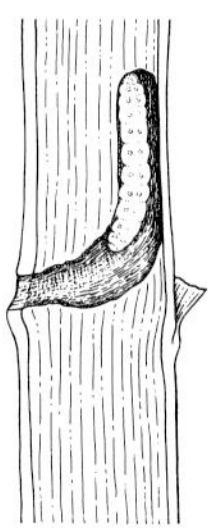

Borers tunneling in stalks, stalks may be broken
European corn borer (p.78)
Southwestern corn borer and related stalk borers (p.101)
Stalk borer (p.106)

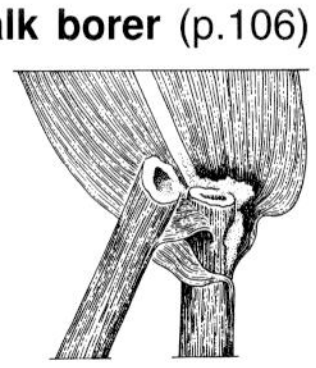

Borers tunneling in base of stalks
Southwestern corn borer and related stalk borers (p.101)

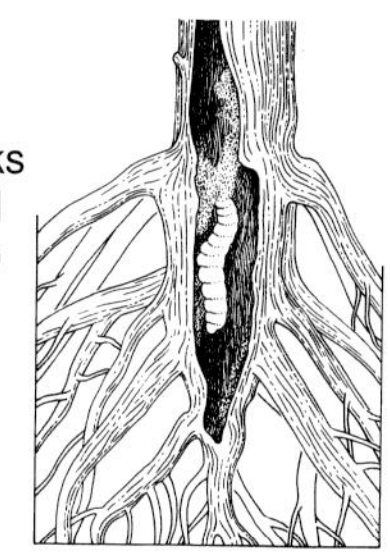

Injury to leaves

Leaves discolored due to feeding by sucking insects

Leaves yellow or red, some may be dying or dead
Corn leaf aphid (p.47)
Greenbug (p.48)
Spider mites (p.104)

Larger leaves with pale stippled areas
Spider mites (p.104)

Silver scars on leaves
Thrips (p.111)

Leaf tissue removed, or blotches in leaves

Blotches in leaves, between upper and lower surfaces
Corn blotch leafminer (p.58)

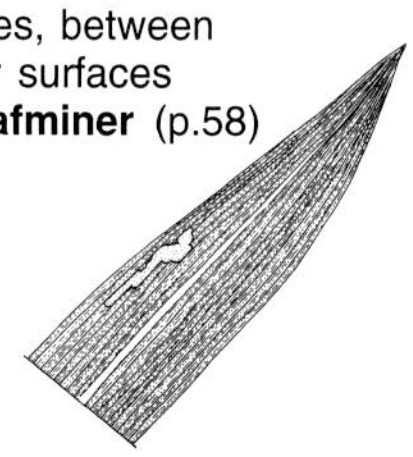

Leaf tissue removed
(Continued on page 28)

(Continued from page 27)

Irregular chunks of leaf tissue eaten along margins
Armyworm (p.50)
Cattail caterpillar (p.55)
Grasshoppers (p.85)
Mormon cricket (p.94)
Yellow woollybear (p.119)

Early- to late-whorl-stage plants with shot-hole injury, small round to oval holes in leaves
European corn borer (p.78)
Southwestern corn borer and related stalk borers (p.101)

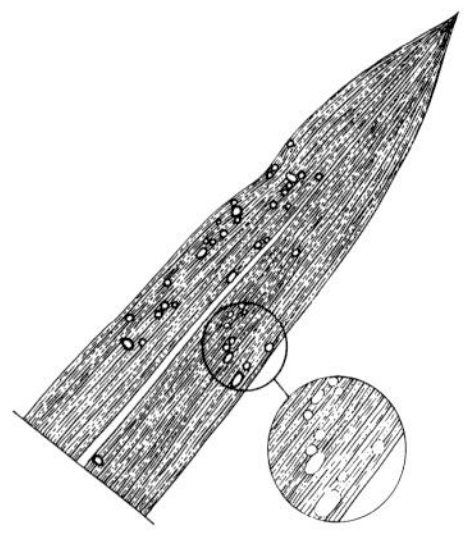

Irregular holes in whorl leaves; injury accompanied by messy frass
Corn earworm (p.59)
Fall armyworm (p.51)
Yellowstriped armyworm (p.52)

Long, irregular strips chewed in mature leaves; sometimes only the epidermis is eaten
Corn rootworm adults (p.61)
Japanese beetle (p.90)

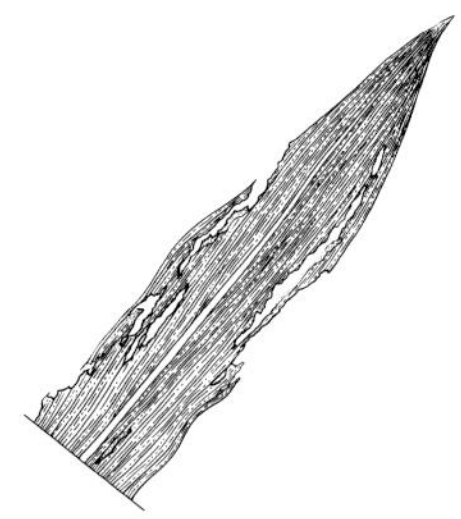

Defoliation so severe that only leaf midribs remain
Armyworm (p.50)
Cattail caterpillar (p.55)
Grasshoppers (p.85)
Mormon cricket (p.94)

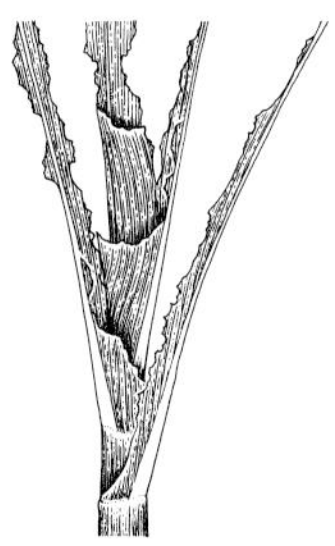

Insect Injury to Corn

Injury to Silks, Ears, or Ear Shanks

Start here

- Injury to ears or silks
  - Kernels chewed upon; may be tunneling in ear or large sections of kernels consumed
    - Ear covered with webbing and frass, kernels chewed upon
      Pink scavenger caterpillar (p.95)

      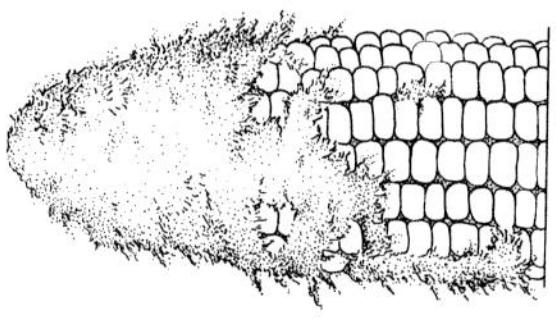

    - Kernels chewed upon, but webbing not associated with injury
      - Injury to ears usually confined to kernels on ear
        - Injury to individual kernels
          - Chewing injury across tops of kernels
            European corn borer (p.78)

            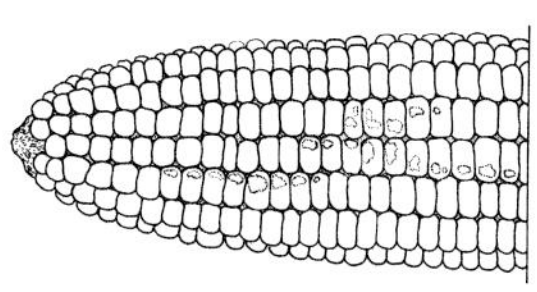

          - Small holes in tops of kernels
            Angoumois grain moth (p.44)

            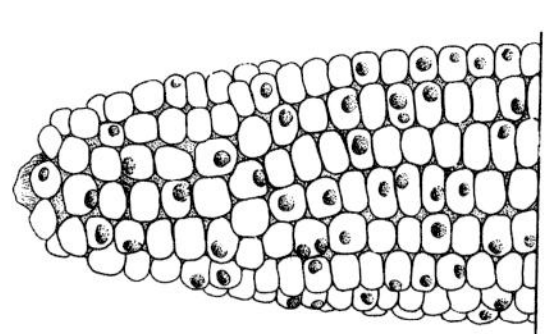

          - Red streaks on kernels
            Wheat curl mite (p.113)
        - Areas on ear (tips, sides) where kernels have been consumed
          Corn earworm (p.59)
          Fall armyworm (p.51)
          Southwestern corn borer (p.101)
          Western bean cutworm (p.113)

          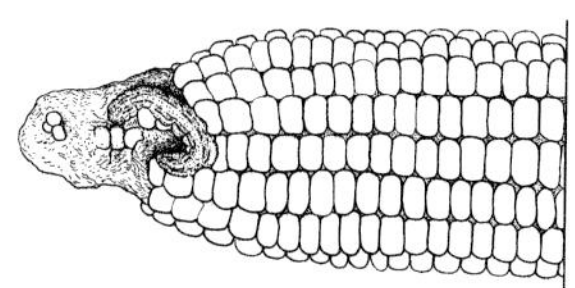

        - Small larvae tunneling or chewing inside kernels in ears
          Sap beetle larvae (p.96)

          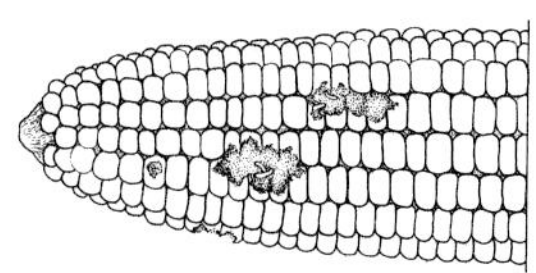

      - Large chunks of ear tissue removed, and large sections of kernels consumed
        Grasshoppers (p.85)

        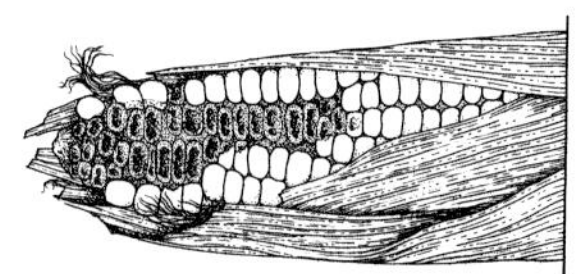

  - Silks clipped or chewed off, but kernels usually not injured; or, if kernels are injured, only the tip kernels are injured
    Corn earworm (p.59)
    Corn rootworm adults (p.61)
    Corn silk beetle (p.68)
    European earwig (p.82)
    Grape colaspis adult (p.85)
    Grasshoppers (p.85)
    Japanese beetle (p.90)
    Western bean cutworm (p.113)
    Yellow woollybear (p.119)

    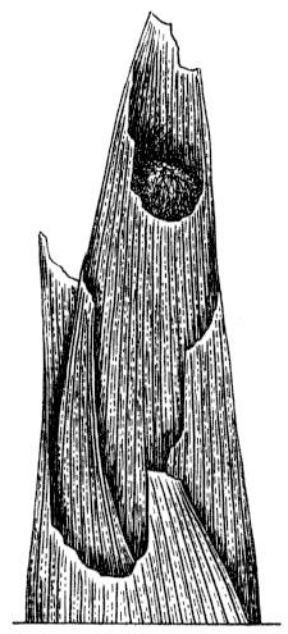

- Injury to ear shanks; shanks may be broken, ears dropped; tunneling inside shank
  European corn borer (p.78)
  Southwestern corn borer (p.101)

  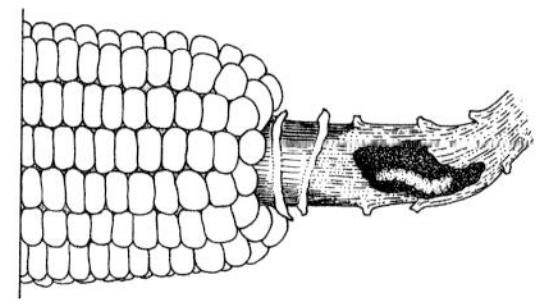

Identification of Insects

The illustrations on this page show certain body regions and anatomical structures of a grasshopper (Mormon cricket), beetle, aphid, and caterpillar. The head, thorax, abdomen, jointed legs, wings, and antennae are indicated, where appropriate, on each of the insects. Specialized structures (e.g., cornicles on the aphid, prolegs on the caterpillar) also are indicated. Also illustrated is a generalized insect leg on which the different segments, or "joints," are indicated. Identification of some insects may rely upon features of the femur, tibia, and/or tarsus.

MORMON CRICKET

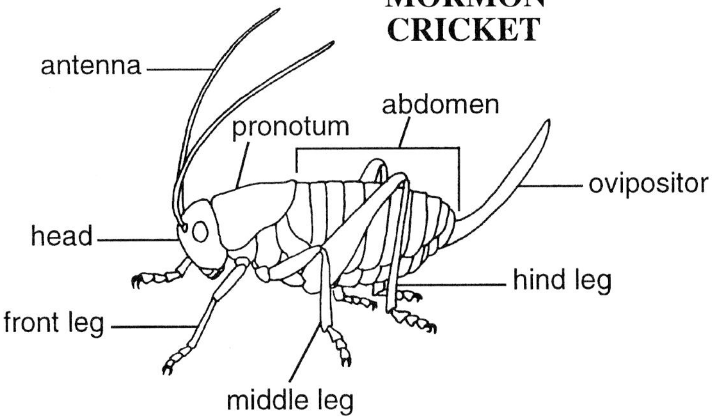

SEGMENTS OF AN INSECT LEG

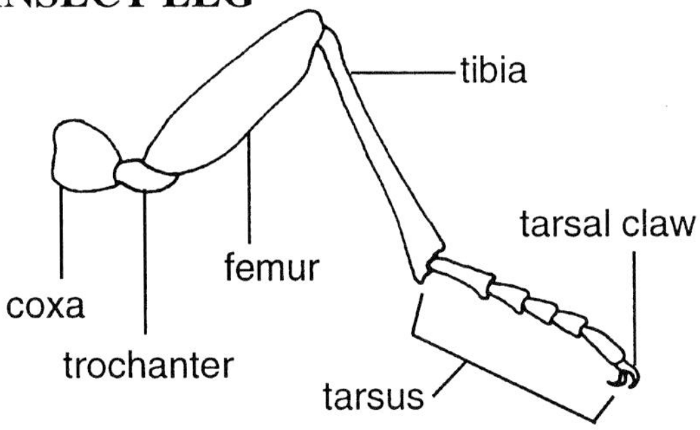

APHID

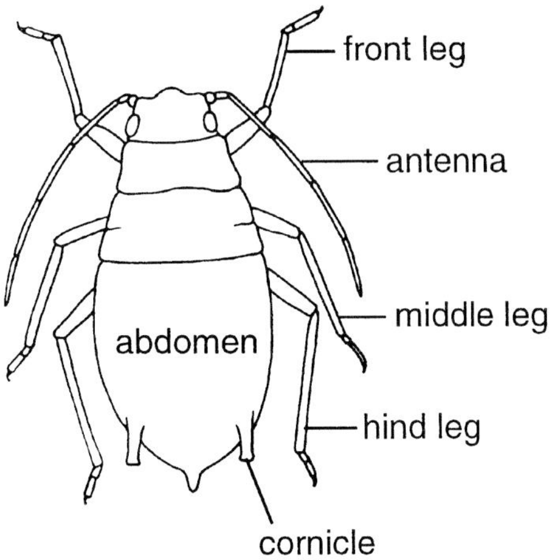

BEETLE

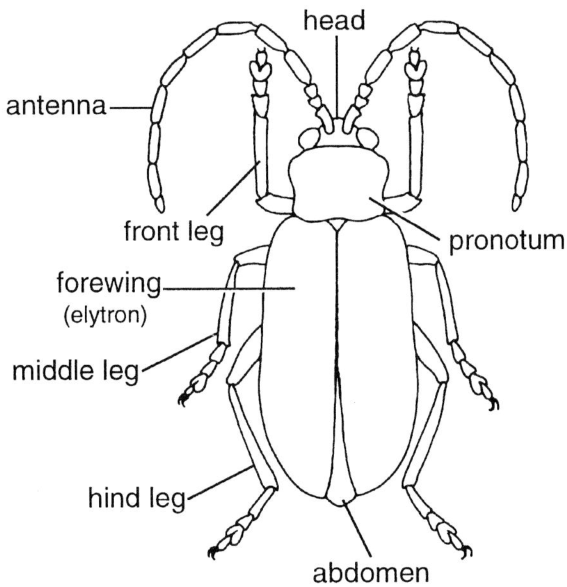

CATERPILLAR

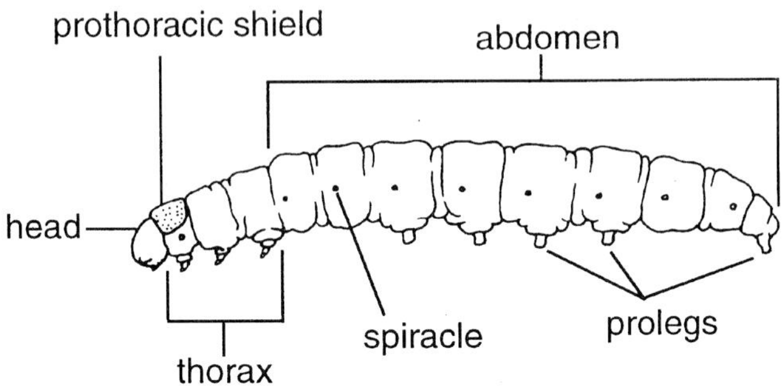

Corn Insect Pests

Pests Feeding Mostly At or Below the Soil Surface

Start here

Legs present

Legs absent
(Continued on page 32)

More than 3 pairs of jointed legs directly behind head

3 pairs of jointed legs directly behind head

12 pairs of jointed legs; body white; max. 6 mm
Garden symphylan (p.110)

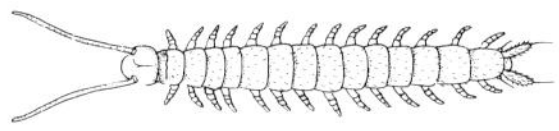

30 or more pairs of jointed legs; body reddish brown
Millipede (p.69)

3 pairs of jointed legs directly behind head, plus 5 pairs of fleshy prolegs on abdomen

3 pairs of jointed legs directly behind head; no prolegs on abdomen

Tail segment with black anal plate; head reddish orange or yellowish brown

Tail segment without black anal plate; head various colors
Cutworms (p.68)

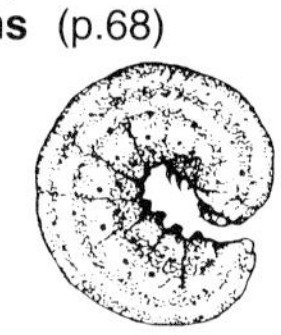

Body long and slender, flattened or slightly C shaped

Body not long and slender, flattened, or C shaped
(Continued on page 33)

Body light to dark pink, shaded with red along the back; Great Lakes area
Potato stem borer (p.88)

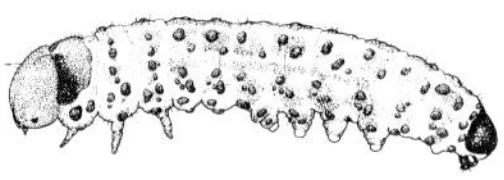

Body dirty white with large brownish purple spots; Iowa, Illinois, Minnesota, Wisconsin
Hop vine borer (p.88)

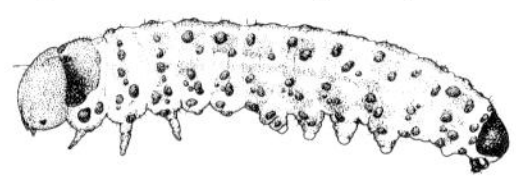

Body C shaped

Body cylindrical or slightly flattened

Dark section on ventral and lateral portions of tail segment

No dark section on ventral and lateral portions of tail segment; bumps on underside of abdomen with clusters of hairs; 3–4 mm
Grape colaspis larva (p.85)

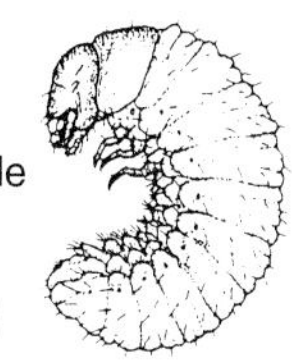

Hairs on ventral side of tail section forming 2 parallel rows, plus scattered hairs
True white grub (p.115)

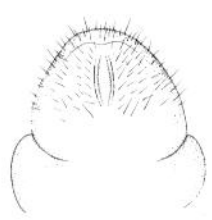

Hairs on ventral side of tail section scattered, not forming a distinct parallel pattern
Annual white grub (p.115)

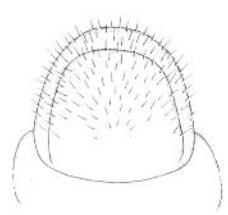

White grub (p.115)

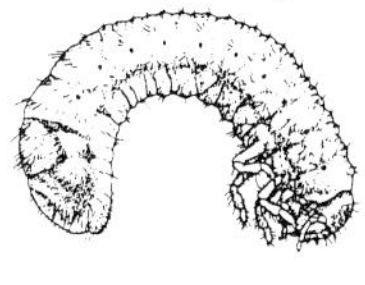

Body soft, white; dorsal side of tail section with dark brown anal plate
Corn rootworm larva (p.61)

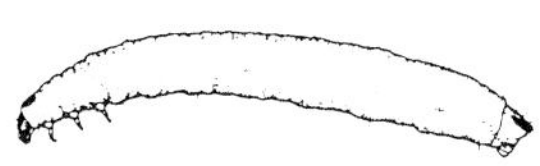

Body hard, yellowish brown, maybe dirty white in mid-section; tail may be slightly flattened, conelike, or with notches
Wireworm (p.117)

(Continued from page 31)

- Head large, not recessed into body; body grublike
  - Abdomen slightly swollen; head dark reddish brown; erect hairs not prominent; 16 mm
    Corn billbug larva (p.53)

  - Abdomen not swollen; head light brown; short, erect hairs prominent; 13 mm
    Whitefringed beetle larva (p.114)

    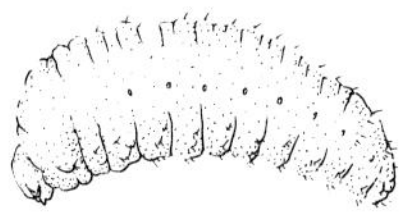
- Head small, often hidden or recessed; body cylindrical and tapered to a point
  - Body dirty white; 6 mm or less
    Seedcorn maggot (p.98)

    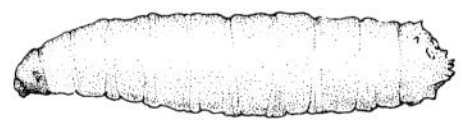
  - Body dark, soil color; larger end with 4–6 fleshy "horns"; often more than 10 mm long
    Crane fly larva (p.69)

    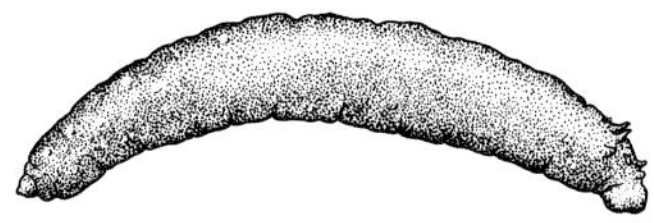

(Continued from page 31)

Body with hard wing covers

More than 12 mm long

Black, dome shaped beetle
Sugarcane beetle (p.109)

Reddish brown, dome shaped beetle
Carrot beetle (p.54)

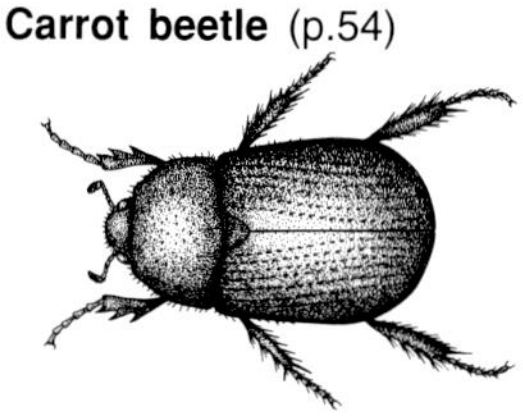

Less than 8 mm long

Body very slender; reddish brown
Slender seedcorn beetle (p.97)

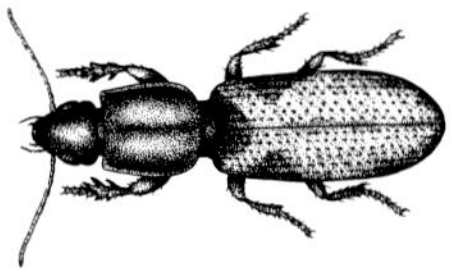

Body not slender, slightly rounded; head and large spots on wing covers black, rest of body yellowish brown
Seedcorn beetle (p.97)

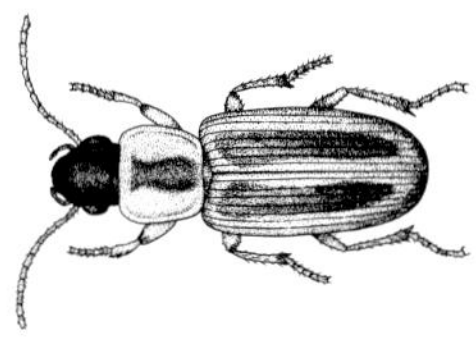

Body lacking wings, or wings not covered

Body oval; pale blue or green; may be covered with fine, white powder; winged forms are black; cornicles present; 1.6 mm
Corn root aphid (p.47)

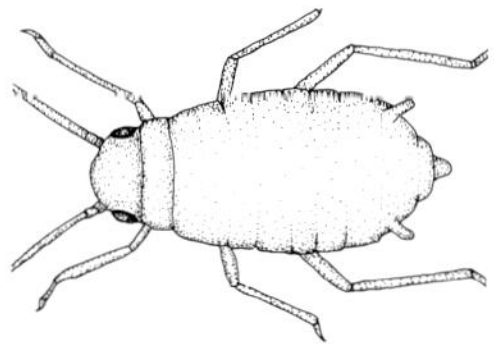

Body antlike; narrowly constricted between thorax and abdomen; dark red or brown

Body with 2 knoblike segments between thorax and swollen section of the abdomen

Dark brown to tan; orange band on top of first abdominal segment; 1.5-5 mm; Gulf Coast states
Red imported fire ant (p.83)

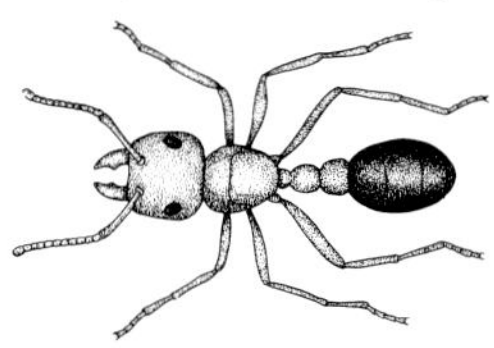

Light yellow to bronze; no orange band on abdomen; 1.3-1.8 mm
Thief ant (p.45)

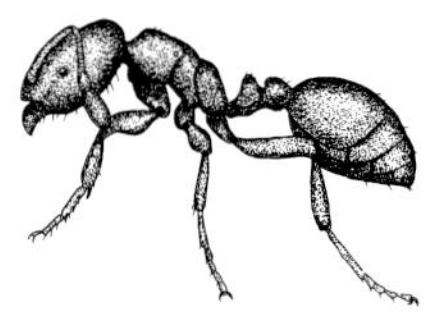

Body with a single reduced segment between thorax and swollen section of the abdomen; 2-2.5 mm
Cornfield ant (p.45)

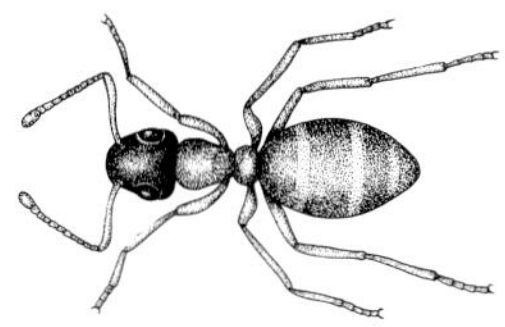

Corn Insect Pests

Pests Feeding Mostly Above the Soil Surface

Start here

Without legs

Snail-like, except lacking a shell
Slug (p.99)

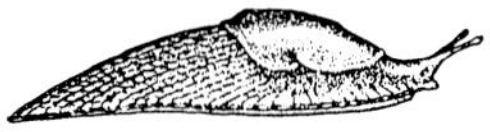

Maggot; pale yellow or white; 1/4 in (6 mm); feeding inside corn leaf
Corn blotch leafminer (p.58)

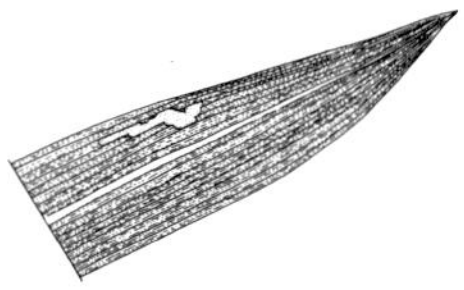

With at least 3 pairs of legs

Microscopic; no larger than a pencil point; yellow to yellowish brown or green; larvae with 6 legs, adults with 8 legs

Body slender, carrot shaped
Wheat curl mite (p.113)

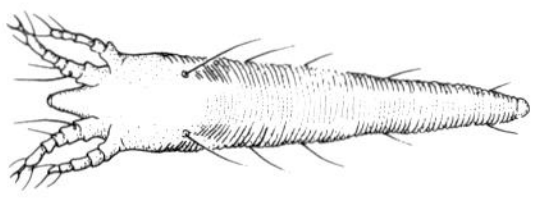

Body oval

Dark areas forming 2 spots on either side of the body; spots not extending to the rear
Twospotted spider mite (p.104)

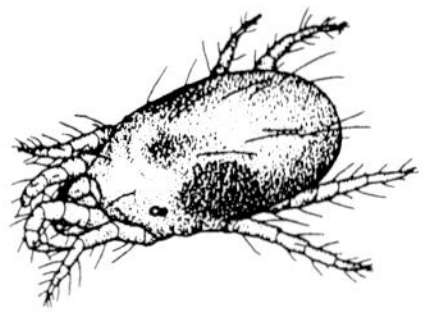

Dark areas of the body extending across the top of the body and to the rear
Banks grass mite (p.104)

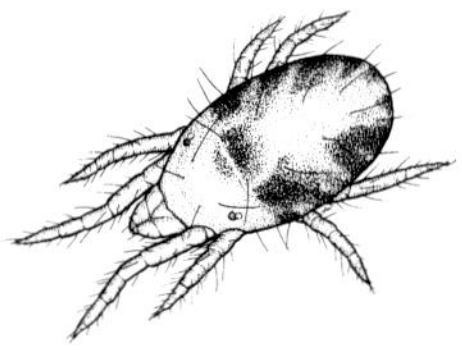

Not microscopic in size

Long, slender hind legs for jumping; femur enlarged; tibia with visible spines

Adults without fully developed wings; female with swordlike ovipositor; Rocky Mountain states
Mormon cricket (p.94)

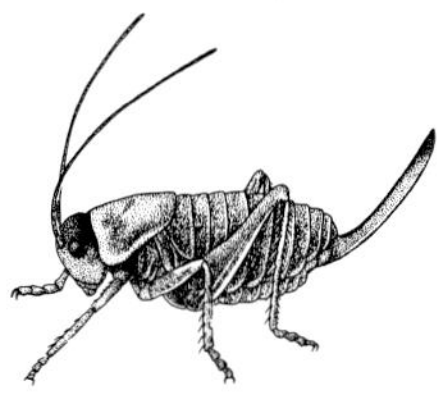

Adults with fully developed wings; females without long, swordlike ovipositor
Grasshoppers (p.85)

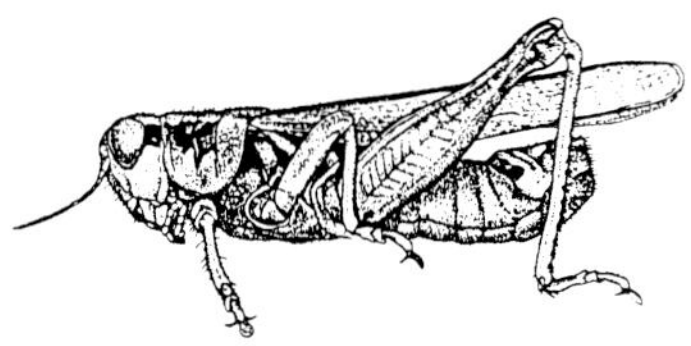

Hind legs, not long and slender; or if long and slender, not used for jumping
(Continued on page 35)

(Continued from page 34)

Body slender and tapered at both ends; very small, 2 mm or less; adults with fringed wings; nymphs without wings
Thrips (p.111)

Adult

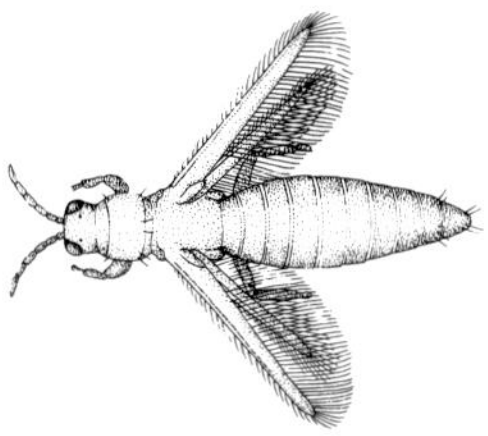

Larva

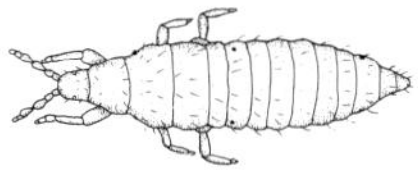

Body not tapered at both ends

Rear of body with pincer shaped structure; 16 mm
European earwig (p.82)

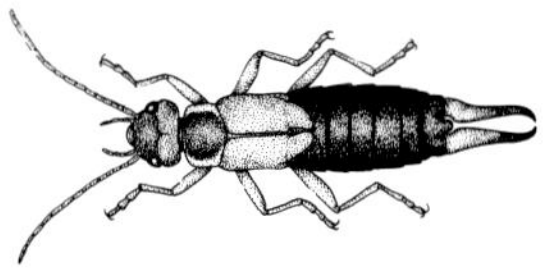

Rear of body without pincer shaped structure

Caterpillar; prolegs on abdomen; body long and often slender
(Continued on page 36)

Beetle, bug, aphid, or leafhopper; body not like a caterpillar; no prolegs on abdomen

Beetle; chewing mouthparts; wing covers present
(Continued on page 38)

Bug, aphid, or leafhopper; piercing-sucking mouthparts, beak very slender and lying under head between front legs (Continued on page 40)

(Continued from page 35)

Body covered with numerous long hairs

Body covered with long dense hairs; white, yellow, or orange
Yellow woollybear (p.119)

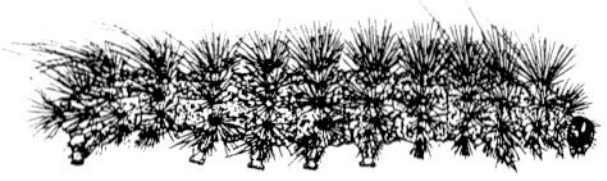

Body sparsely covered with tufts of long, dense hairs; color a mixture of black, orange, and white
Cattail caterpillar (p.55)

Body sparsely covered with long hairs, or hairs not noticeable

Body with lateral stripes, often extending from head to tip of abdomen

No dark stripe on side of head; no dark purple or brown "saddle" behind true legs

Prominent white inverted Y pattern on "face"; 4 dark tubercles on top of each abdominal segment, largest on segment 8
Fall armyworm (p.51)

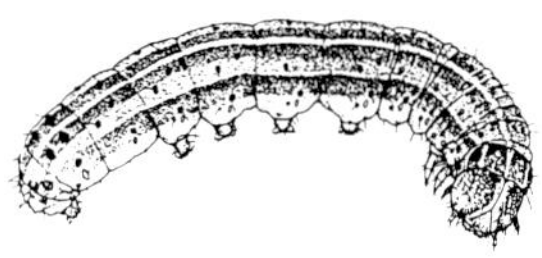

No prominent white inverted Y pattern on "face"

Velvety black with yellow stripes along sides; double row of black triangular markings on back
Yellowstriped armyworm (p.52)

Not velvety black with yellow stripes or double row of black triangles

Skin coarse and covered with microspines; light green to pink to brown with alternating light and dark stripes
Corn earworm (p.59)

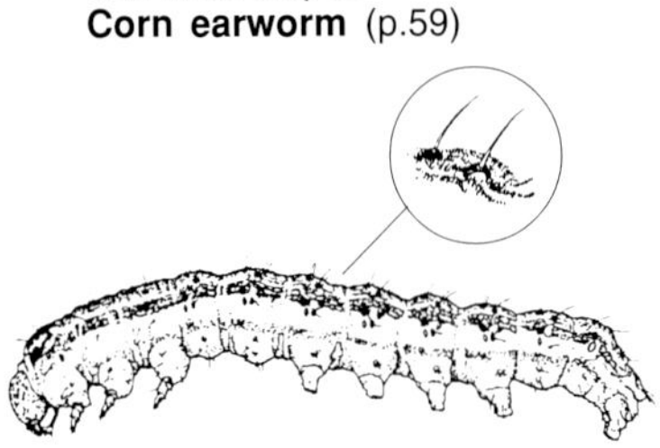

Skin smooth, lacking microspines; 2 crescent shaped dark marks on "face"; prolegs often with dark stripes; body with alternating light and dark stripes
Armyworm (p.50)

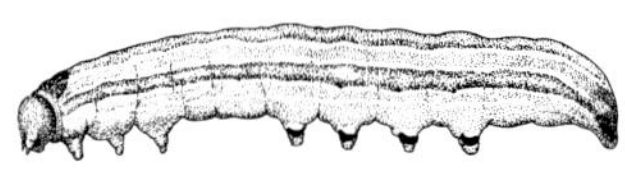

Dark stripe on side of head; young larvae with dark purple "saddle" behind true legs; "saddle" faded in older larvae
Stalk borer (p.106)

Body without stripes along sides
(Continued on page 37)

(Continued from page 36)

Bluish green with brown or purple bands around each body segment
Lesser cornstalk borer (p.94)

Not bluish green with brown or purple bands; with or without spots

Light green or dark green with white spots; on seedling plants, often inside leaves tied together
Leafrollers (p.93)

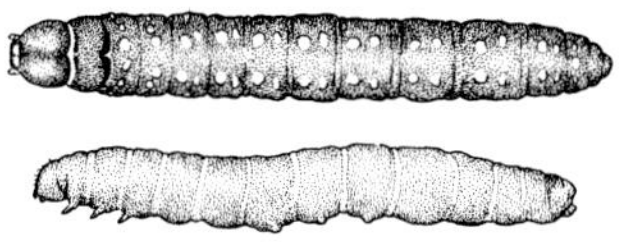

Not green, and without white spots

Lacking dark spots on abdomen

Dark spots on abdomen

Body and head bright pink; pronotum ("neck") dark; feeding on kernels in ear
Pink scavenger caterpillar (p.95)

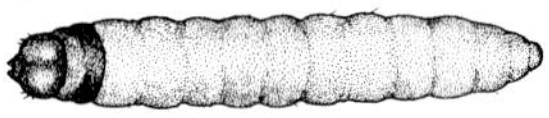

Body and head not bright pink, rather light brown or white

Body light or dark, without large black spots

Body white with large black spots; head orange
Southwestern corn borer (summer phase) (p.101)

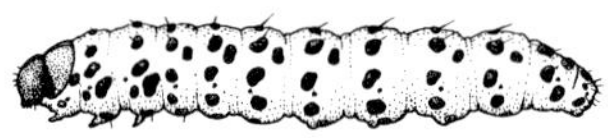

Body white; head only slightly darker

Body light brown; head dark; pronotum ("neck") with large, dark spots; usually in corn ear
Western bean cutworm (p.113)

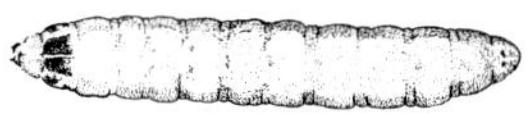

Each abdominal segment with 6 large, shiny tubercles
Sod webworm (p.112)

Each abdominal segment with small tubercles appearing as dots or halos
European corn borer (p.78)

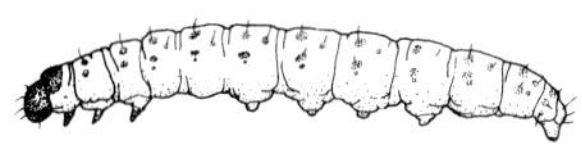

Mature larva 6 mm long; in grain kernel
Angoumois grain moth larva (p.44)

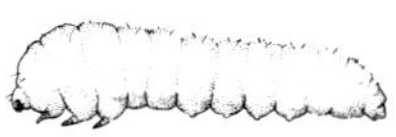

Mature larva 25 mm long; in corn stalk
Southwestern corn borer (winter phase) (p.101)

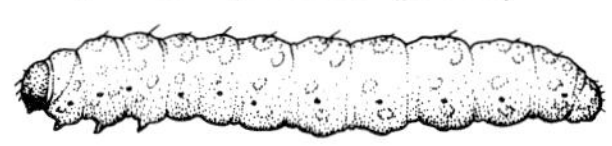

Identification and Diagnosis

(Continued from page 35)

- Weevils; head with snout that is very long or short and broad
  - Snout short and broad; wing covers with numerous short, stiff hairs; 10–12 mm **Whitefringed beetle** (p.114)

    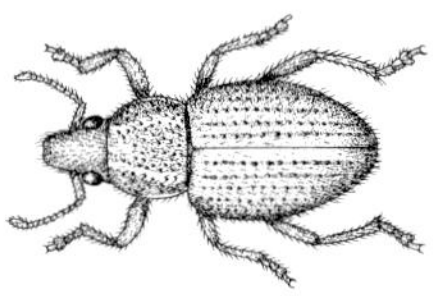
  - Snout long; wing covers without numerous short, stiff hairs; 7–13 mm **Billbugs** (p.53)

    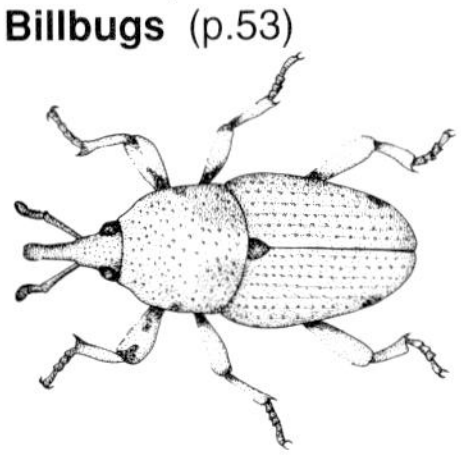
- Head without snout
  - Wing covers not extending to tip of abdomen
    - Wing covers without 4 yellow spots; 3-4 mm **Corn sap beetle, Dusky sap beetle** (p.96)

      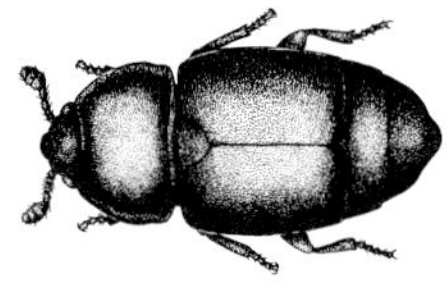
    - Wing covers with 4 yellow spots; 8 mm **"Picnic beetle"** (p.96)

      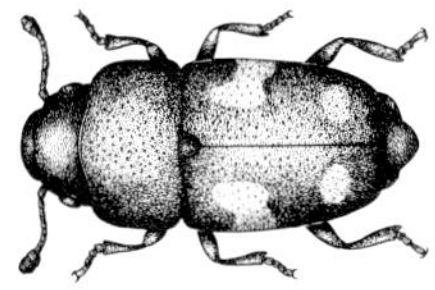
  - Wing covers extending to tip of abdomen
    - Less than 8 mm long
      - Body color dark blue or black, maybe shiny metallic
        - More than 2 mm long
          - Wing covers shiny blue or bluish green **Cereal leaf beetle** (p.56)

            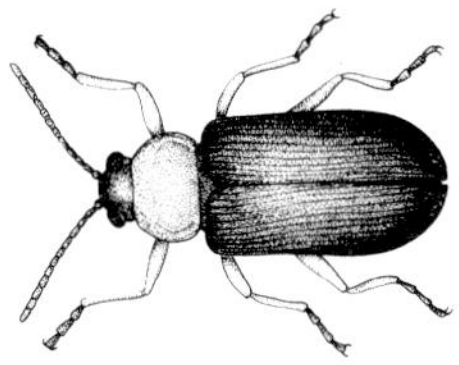
          - Wing covers black
            - Wing covers dull black and densely covered with small punctures; pronotum with small lateral projections; 5 mm **Southern corn leaf beetle** (p.100)

              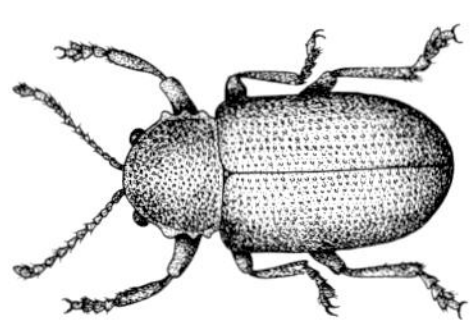
            - Wing covers shiny black without visible punctures; pronotum red or orange; 3-4.5 mm **Corn silk beetle** (p.68)

              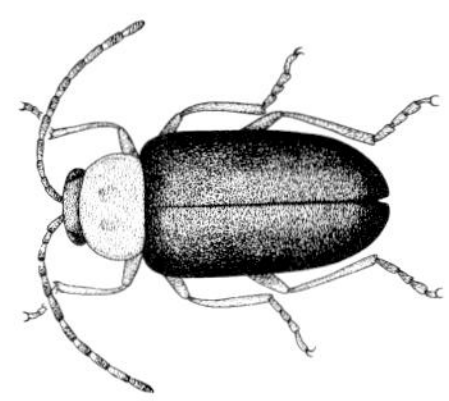
        - 1.5 mm long, size of a pinhead; metallic black; hind femur enlarged **Corn flea beetle** (p.84)

          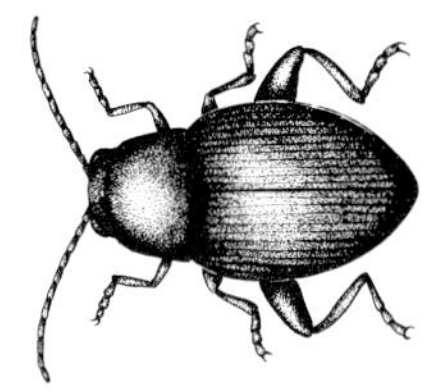
      - Body color not totally dark or shiny metallic (Continued on page 39)
    - More than 8 mm long; body metallic green; wing covers bronze; 2 white spots at tip of abdomen **Japanese beetle** (90)

P/U FROM SOYBEAN BOOK p.28

(Continued from page 38)

Wing covers with 12 black, oval spots
Spotted cucumber beetle (p.63)

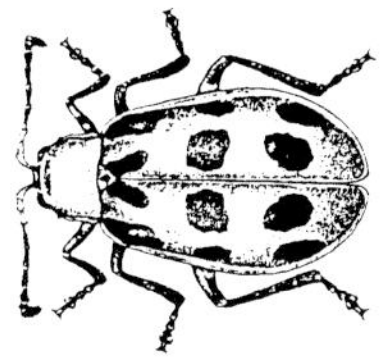

Wing covers without black oval spots

Wing covers with 1 wide, black stripe (male) or 3 narrow black stripes (female)
Western corn rootworm (p.64)

Male

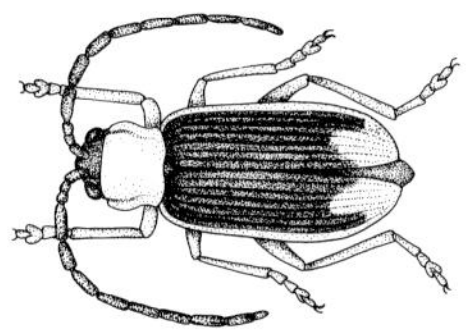

Female

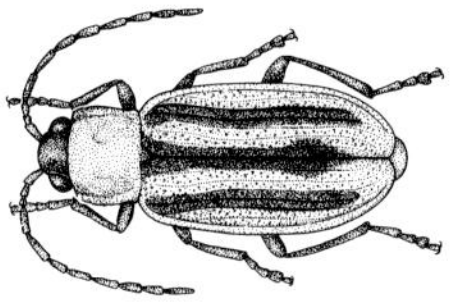

Wing covers tan, green, or yellow, and without black stripes

Wing covers with numerous dark pinstripes and punctures
Grape colaspis (p.85)

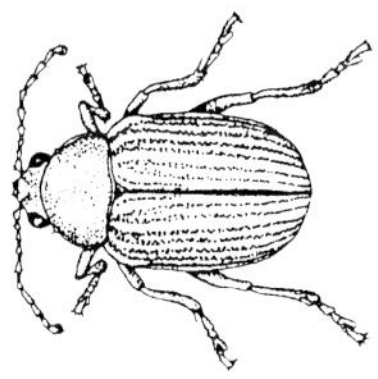

Wings without numerous dark pinstripes or punctures

Legs black or pale with a mixture of black
Mexican corn rootworm (p.61)

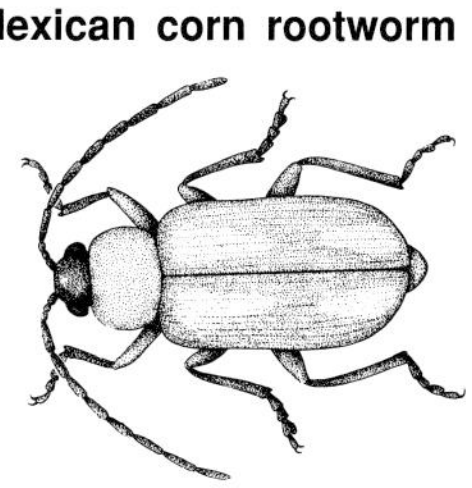

Legs yellow or green, not black
Northern corn rootworm (p.62)

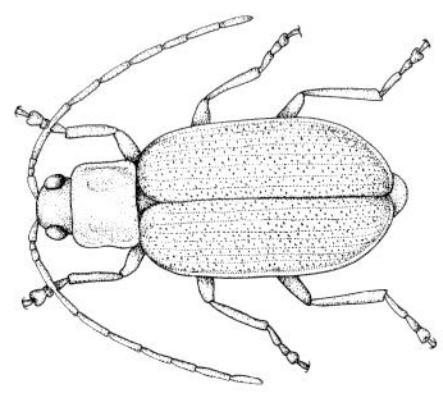

(Continued from page 35)

No pair of tubelike structures (cornicles) on rear of abdomen

Body oval; 2 tubelike structures (cornicles) on rear of abdomen; wings clear on winged forms; not very active, often in colonies
Aphids (Continued on page 41)

Insect not in spittle mass on plant

Insect in spittle mass on plant; body light green with red eyes
Twolined spittlebug (p.111)

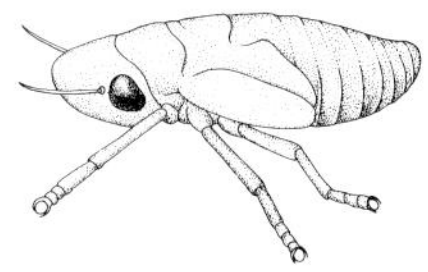

Wings very short, not extending more than half the length of the body; abdomen swollen; 2.5-3.5 mm
Corn delphacid (p.58)

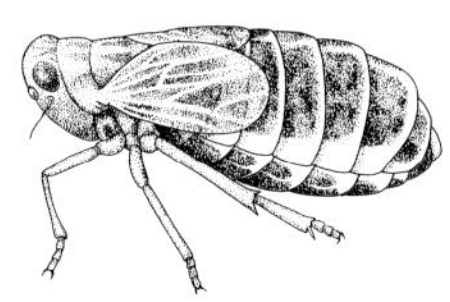

Wings nearly as long as body, or longer

Body dark gray, brown, or pale green

Body black with white wings; 2 black spots on either side of wings; wings overlapping edge of abdomen; 5 mm
Chinch bug (56)

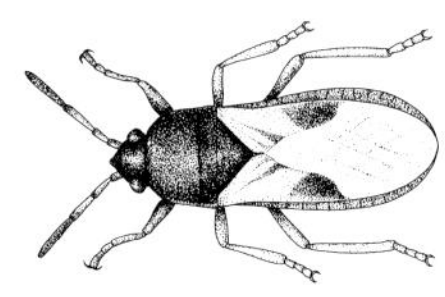

Body gray; wings transparent and covering all of abdomen; 3-4 mm
False chinch bug (p.82)

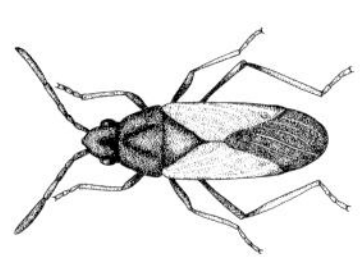

Body brown and shield shaped; 10-15 mm
Stink bugs (p.108)

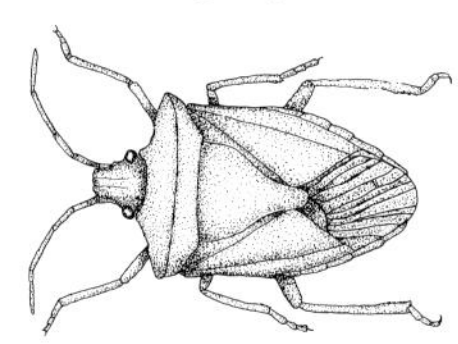

Body pale green and wedge shaped; black spots between eyes

2 large, round, black spots between eyes; 3-5 mm
Corn leafhopper (p.91)

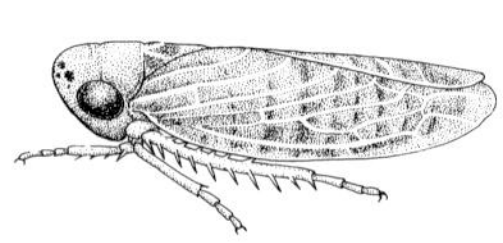

4 small, round, black spots between eyes; 3-4 mm
Blackfaced leafhopper (p.91)

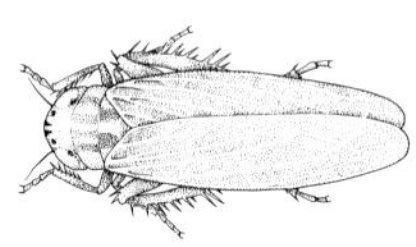

(Continued from page 40)

Body black, some individuals may have white, waxy pubescence
Bean aphid (p.46)

No illustration

Body not black, often light green to dark green or bluish green

Cornicles long and slender, reaching almost to the tip of the abdomen; body either green or pink
Potato aphid (p.49)

No illustration

Cornicles short, rarely reaching the tip of the abdomen

Body pale green with narrow darker green stripe along the center of the abdomen
Greenbug (p.48)

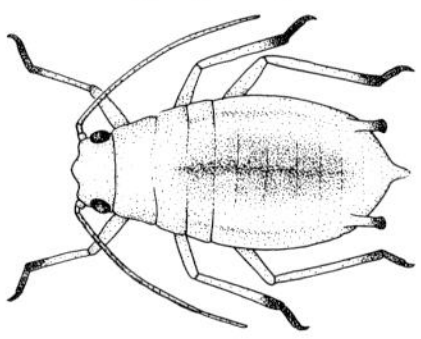

Body dark green or bluish green

Body dark green; base of cornicles red or orange
Bird cherry-oat aphid (p.46)

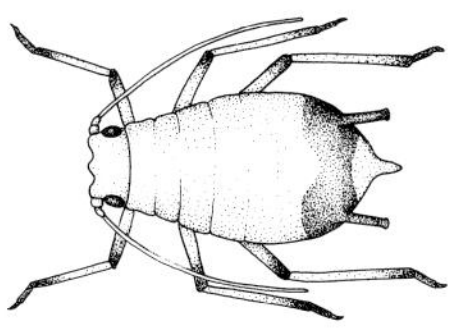

Body dark bluish green; base of cornicles black
Corn leaf aphid (p.47)

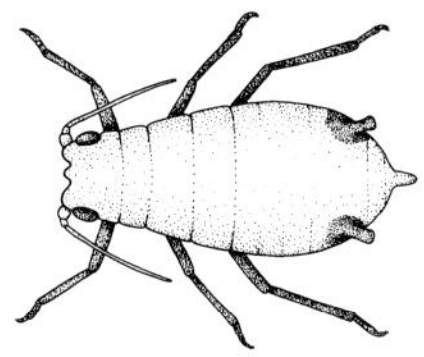

Parasitoids and Predators of Insect Pests of Corn

A key to the many species of parasitoids and predators would be impractical for this handbook. However, illustrations on this page show representations of each group of these natural enemies. These illustrations can be used for insects that do not match any of the illustrations in the key.

Tachinid Flies, (p.125). Adults are 1/5–1/2 in (4–12 mm) long, depending upon species. They are relatively robust and bristly, and beelike or wasplike in appearance.

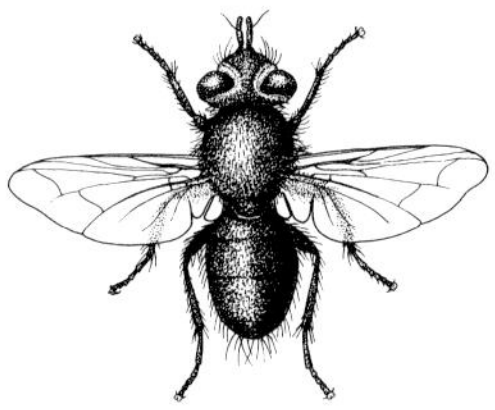

Braconid Wasps, (p.126). Braconids are small, usually dark wasps about 3/8 in (10 mm) or less in length, with long, curved antennae. Females of some species have very long ovipositors.

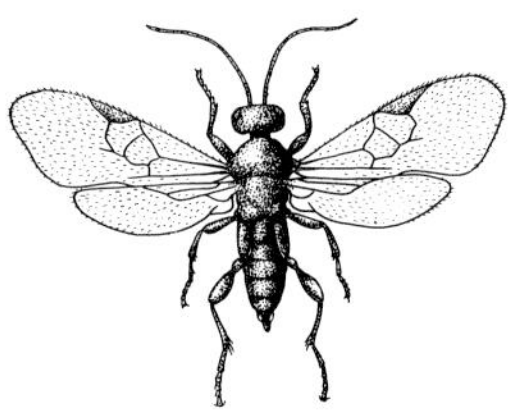

Ichneumon Wasps, (p.127). Adults are slender and vary from black to yellow-brown. They have long antennae with 16 or more segments. The female's ovipositor is permanently extended and usually is as long as or longer than the body. Adults vary from less than 1/8 in (3 mm) to 1 1/2 in (38 mm) long, not including the antennae and ovipositor.

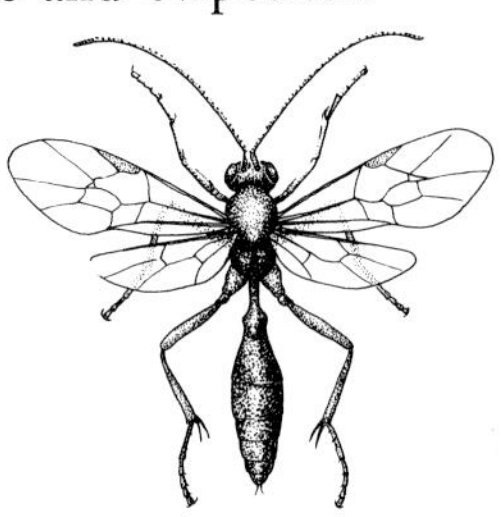

Trichogrammatid Wasps, (p.128). Adult *Trichogramma* wasps are very tiny (less than 1/32 in [0.7 mm] long), look like tiny ants, and are difficult to see. They have microscopic hairs on the wings, usually arranged in rows.

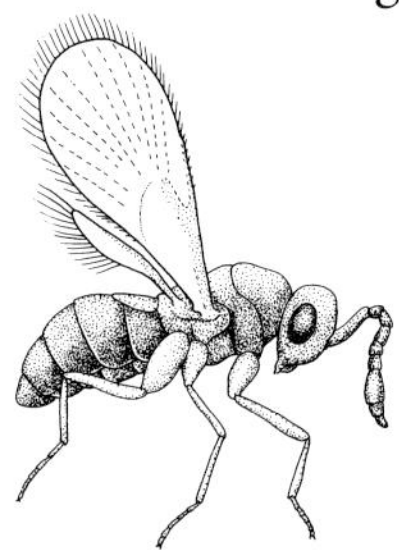

Ground Beetles, (p.130). Ground beetle adults and larvae have chewing mouthparts. (a) The adults vary considerably in size, shape, and color. Adults range in length from less than 1/4 in (6 mm) to more than 1 in (2.5 cm). Most species are dark, shiny, and somewhat flattened, ranging in shape from round to ovoid to having well-defined head, thoracic, and abdominal regions. Many species have striate wing covers (wing covers with grooved or depressed lines). (b) The larvae usually are distinctly segmented and are tapered toward the rear. They have large heads with pronounced mandibles, and they have well developed legs.

a.

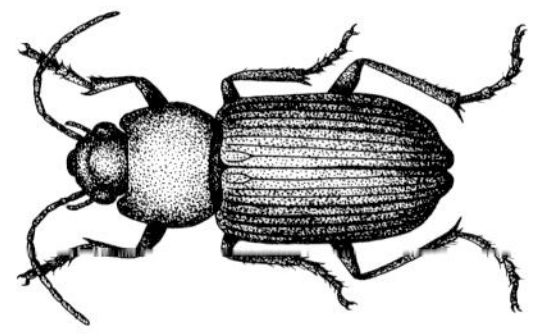

b.

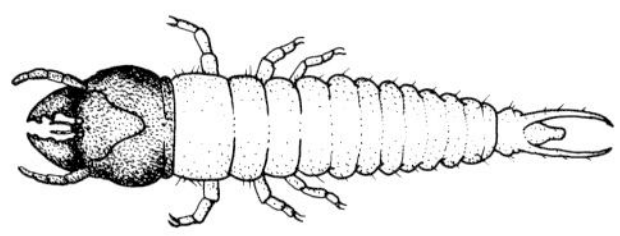

Rove Beetles, (p.130). Rove beetles are slender and elongate black or brown insects about 1/2 in (10 mm) or less in length. The beetles have very short wing covers, so much of their abdomen is exposed. Their mandibles (chewing mouthparts) are long, slender, and sharp, and usually cross in front of the head.

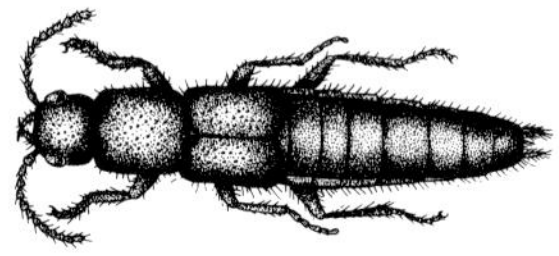

Lady Beetles, (p. 131). Lady beetle larvae and adults have chewing mouthparts. (a) Lady beetle larvae often are black, brown, or dark blue, with orange or yellow stripes across the back. The larva's body, which often is spiny, is slender and tapered with well-developed legs. (b) Pupae typically are orange to dark brown, somewhat C-shaped, and attached by the tip of the abdomen to a leaf or stem. (c) Adults are oval, convex, and often brightly colored (orange or red) with black spots or markings. Their tarsi are 3-segmented.

a.
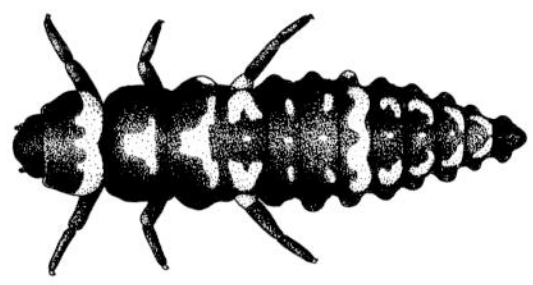

c.

b.
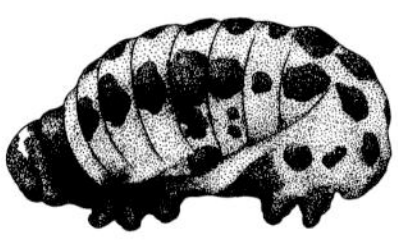

Flower Bugs and Minute Pirate Bugs, (p.133). Adult *Orius* are 1/8 in (2–3 mm) long and appear black and white (black thorax and amber-clear wings). Their wings are thickened and leathery near the thorax and membranous near the tip of the abdomen, and the membranous tips overlap. They have piercing-sucking mouthparts.

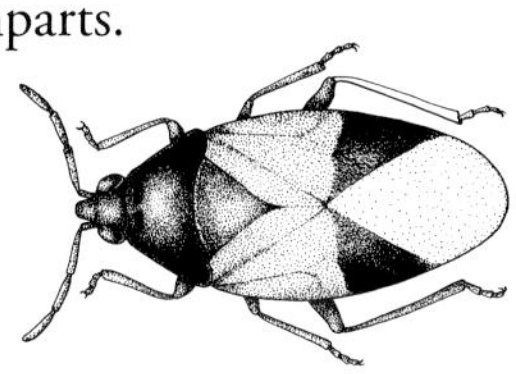

Damsel Bugs, (p.135). Adult damsel bugs are 1/5–1/2 in (5–12 mm) long and slender, with slightly enlarged front legs. Most common damsel bugs are either mottled brown, pale yellow, or black. Both long- and short-winged forms occur. Their wings are thickened and leathery near the thorax and membranous near the tip of the abdomen, and the membranous tips overlap. They have piercing-sucking mouthparts.

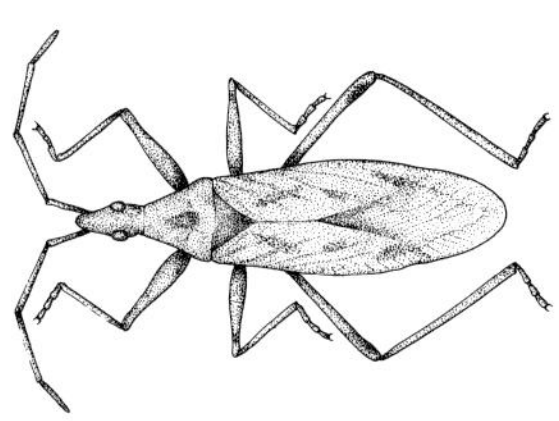

Predaceous Stink Bugs, (p.135). Predaceous stink bugs are broad, shield-shaped, and 1/3–3/4 in (7–20 mm) long. Most species are brown on top and paler below, although some are brightly colored. Predaceous species have thicker, stronger front legs and thicker, stouter mouthparts (piercing-sucking) than plant-feeding species.

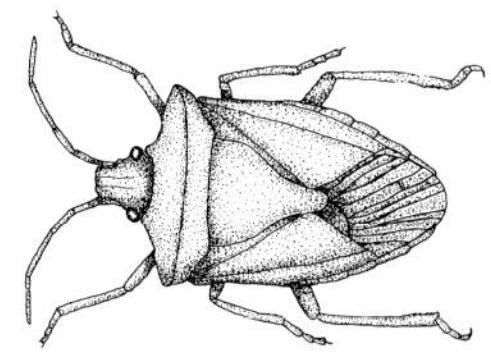

Lacewings, (p.136). Lacewing adults and larvae have chewing mouthparts. (a) Larvae are elongate and flat and have large sickle-shaped mandibles that they use to pierce prey and suck body fluids. (b) Adult green lacewings are about 1 in (25 mm) long and generally are green or yellow, although a few species are dark colored. Adult brown lacewings are brown or yellow-brown and generally are smaller than green lacewings. Both green and brown lacewing adults have large, lacy wings. Green lacewings have threadlike antennae; brown lacewings have antennae that resemble strings of beads.

a.
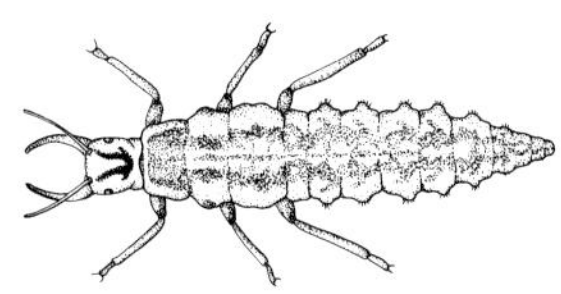

b.
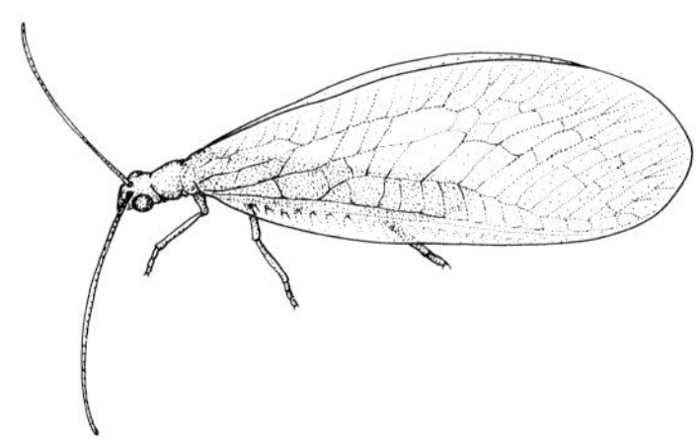

Pest Information

For each pest, the following information is provided. Although the information is presented in a uniform format, more information is provided for pests that are more common or more economically important in corn.

Scientific Classification. The genus and species of an insect pest is followed by order and family names (family name is in parentheses). Some deviations from this format occur for pests that are not insects.

Origin and Distribution. This section indicates whether the pest species is native or introduced and its approximate distribution in North America.

Description. Identification details are provided for the stages that injure corn, and sometimes for noninjurious stages such as eggs and pupae.

Pest Status. This section explains the economic importance of the insect throughout its range.

Injury. The injury caused by the pest is described. If more than one stage of the pest causes injury to corn or if the pest causes injury at more than one stage of corn growth, all types of injury are described.

Life History. Details about a pest's life cycle are provided, including, but not limited to, overwintering stage(s), number of generations per year, time of occurrence and location of all stages, and developmental times.

Management. All potential and practical tactics for management of the pest are explained. When the information is known, the impact of natural biological control by beneficial organisms is discussed. Widely used economic thresholds or other decision-making tools are identified. Specific insecticides are not listed.

Selected References. The numbers of selected references at the end of most sections refer the reader to the list of references near the back of the handbook.

Distribution Map. The distribution of the pest in North America is indicated on the map. If the pest occurs throughout North America wherever corn is grown, or if the distribution of the insect is not well known, a map is not provided. These distributions are approximations for most insects; some pests occur outside of the range indicated on the map. In particular, information at the southern and northern limits of a species' range usually is uncertain.

Photographs. Color photographs of the injurious stages of the pest and the injury caused by the pest are included for most species. However, photographs for injury caused by some of the pests were not available. Photographs of noninjurious stages of some of the most economically important pests also are included.

Angoumois Grain Moth

Angoumois grain moth larvae (Marlin E. Rice).

Scientific Classification. *Sitotroga cerealella*, Lepidoptera (Gelechiidae). Related species are the almond moth, *Cadra cautella*, and the Indianmeal moth, *Plodia interpunctella*, both Lepidoptera (Pyralidae).

Origin and Distribution. The Angoumois grain moth probably originated in Europe. It is found throughout North and Central America, most commonly in the southern United States.

Description. The full-grown larva, found inside corn kernels, is about 1/5 in (5 mm) long and has a white body, yellow head, three pairs of true legs, and four pairs of prolegs on the abdomen. Adults, often found on the surface or in the top 1 ft (1/3 m) of stored corn, are 1/5–1/4 in (5–7 mm) long and pale brown with narrow, sharply pointed fringed wings (wingspan of 1/4–5/16 in [7–9 mm]).

Although both the almond moth and Indianmeal moth are found in stored corn, these moths' wings differ from those of the Angoumois grain moth. The wings of the Angoumois grain moth are buff to gray, and the tips of the hind wings bear a fingerlike projection. The wings of the almond moth and Indianmeal moth are patterned, and the hind wings do not bear fingerlike projections. Additionally, almond moth and Indianmeal moth larvae spin webbing on the surface of stored corn; Angoumois grain moth larvae do not produce webbing.

Pest Status. The Angoumois grain moth is the most serious caterpillar pest of stored corn because it infests both intact and damaged kernels. It is a problem primarily in the southern United States.

Injury. The larva feeds inside and hollows out a kernel of corn, making it unfit for consumption or planting. The adults are short lived and do not feed.

Life History. Breeding may be continuous throughout the year in warm storage facilities. Females lay eggs on the surface of kernels both in the field and in storage. After hatching, larvae penetrate into kernels and complete development within the kernels. Adults emerge from the kernels and mate. Females lay 200–300 eggs during a 1–2-week life span. The life cycle is completed in less than 4 weeks at 85°F (30°C) but is prolonged at cooler temperatures. Larvae may be parasitized by the wasp *Pteromalus cerealellae*.

Management. Although females may lay eggs on corn kernels in the field, management of infestations in the field is not recommended. Management tactics for Angoumois grain moth in stored corn are similar to those for *Sitophilus* weevils. Grain storage facilities should be cleaned thoroughly before new corn is stored. Chemical protectants may be applied during loading of the bin. Aeration to cool the corn helps suppress population increase.

Corn that has been treated properly and aerated periodically usually does not become infested. However, pheromone traps placed within the bin can be used to monitor for adult males. If infestations become established, fumigation may be necessary. The status of pesticide resistance in this moth is uncertain, but resistance to malathion has been reported.

Selected Reference. 120

J. E. Baker and J. E. Throne

Ants (Cornfield and Thief)

Scientific Classification. Cornfield ant, *Lasius alienus,* formerly *Lasius niger* var. *americanus* and *L. americanus*; thief ant, *Solenopsis molesta*, Hymenoptera (Formicidae).

Origin and Distribution. Cornfield and thief ants are distributed widely throughout the corn-growing areas of North America. The cornfield ant is native to both North America and Eurasia; the thief ant is native to North America.

Description. The cornfield ant worker is 1/12–1/10 in (2–2.5 mm) long, brown to very dark, and has a one-segmented abdominal "waist," or pedicel. The antennal segment closest to the head is slightly shorter than the head, and a fringe of hair encircles the tip of the abdomen. The thief ant worker is 1/20–1/14 in (1.3–1.8 mm) long, yellow to brown, and has a two-segmented abdominal pedicel. The antenna has an elongated, two-segmented club on the end, and the antennal segment closest to the head is half the length of the head.

Cornfield ant and corn root aphids (Marlin E. Rice).

Pest Status. Cornfield and thief ants are sporadic pests of corn. The cornfield ant cultivates corn root aphids that feed on corn roots. Infestations are particularly troublesome in no-till or in reduced tillage continuous corn. The thief ant preys on larger ant species but also attacks planted seeds and germinating corn.

Injury. The association between cornfield ants and corn root aphids may result in injury to corn plants. Feeding by the aphids may retard the growth of the plants, especially during dry years. Leaves of infested plants become yellow or red, then turn brown and die. Small anthills usually are

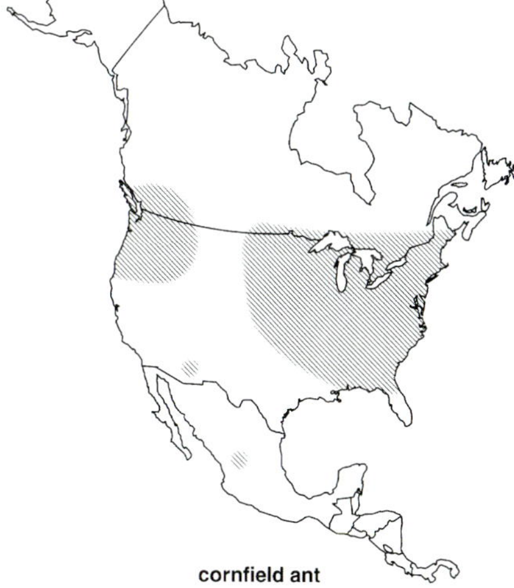

found around injured plants; the small brown ants may be found tunneling along the corn roots. Plants rarely are killed.

Infestations of thief ants are uncommon but can result in reduction of plant population. The ants scatter little grains of starch about an empty seed after hollowing out the contents.

Life History. Cornfield ant males and winged females are common in the nest from midsummer to fall. Females begin new colonies in the fall; however, egg laying may not begin until spring. Workers first emerge in July then collect eggs of corn root aphids in the fall and store them in their nests during the winter. In early spring, cornfield ants carry newly hatched aphids to roots of suitable weed hosts, particularly smartweed and dock. When corn is available, the ants transfer the aphids to corn roots. The ants derive a sticky excretion called honeydew from the aphids. A disease caused by the fungus *Laboulbenia formicarum* may infect cornfield ants.

Thief ants usually live in association with nests of larger ant species. They enter the nests through narrow passages to kill and eat the immature stages of the larger ants. Although they are predaceous, they also feed on planted seeds of corn, milo, sorghum, and other grains.

Management. Crop rotation reduces the potential for an increase in populations of cornfield ants and corn root aphids. Control of weed hosts early in the spring eliminates feeding sites for corn root aphids. Tillage operations weaken colonies of both ant species and reduce aphid survival. When replanting is necessary because of reduced plant populations, a soil insecticide can be used to protect the new stand.

Selected References. 105, 173, 225

Ricardo Bessin

Aphids

Six species of aphids infest corn in North America. Because each species poses a different threat, they are discussed individually. "Management of Aphids" applies to all species.

Bean Aphid

Scientific Classification. *Aphis fabae*, Homoptera (Aphididae).

Origin and Distribution. The bean aphid is widely distributed in the Northern Hemisphere. Although found in some parts of South America and Africa, this aphid is not common in the tropics.

Description. Bean aphids are small to medium-sized black insects that are 1/16–1/8 in (1.6–3.1 mm) long. A white, waxy pubescence may develop on individuals in older colonies.

Pest Status. Bean aphids are a minor pest of corn and are more prevalent in Europe than in the United States.

Injury. Bean aphids are potential vectors of corn viruses such as maize dwarf mosaic virus (MDMV) and barley yellow dwarf virus.

Life History. Most of the information about the biology of bean aphids comes from Europe. European *Euonymus* and European cranberrybush *Viburnum* are the primary overwintering hosts. In the spring, winged aphids (alates) migrate and establish colonies on many secondary hosts, including corn. Winged aphids return to the primary hosts in the fall where they reproduce sexually.

J. W. Bing

Bird Cherry-Oat Aphid

Scientific Classification. *Rhopalosiphum padi*, formerly mistaken for *R. fitchii* and *R. prunifoliae*, Homoptera (Aphididae).

Origin and Distribution. The bird cherry-oat aphid is distributed worldwide.

Description. Wingless (apterous) forms of the bird cherry-oat aphid on corn usually are dark olive to green-black, with orange, red, or rusty-color patches around the bases of the cornicles, or tubelike structures on the rear of the abdomen. Head and thorax of winged forms are mostly black; abdomen is light green with darker green markings on the sides. Winged bird cherry-oat aphids are difficult to distinguish from the corn leaf aphid. Winged and wingless adults average 1/16–3/32 in (1.6–2.4 mm) long.

Pest Status. Populations of the bird cherry-oat aphid develop on corn from July until frost almost every year in the western Corn Belt. These aphids seldom occur on tassels but are found in mixed colonies with the corn leaf aphid on ear husks and leaf surfaces. Large colonies may form on the lower stalk.

Bird cherry-oat aphids (Marlin E. Rice).

Injury. Significant loss of corn yield caused by the bird cherry-oat aphid has not been documented. However, it may transmit strains of barley yellow dwarf virus, first to corn and later in the season to winter wheat or other grasses. Although the effects of barley yellow dwarf virus in field corn are largely innocuous, in some areas of the country corn may serve as a reservoir for the virus, thus facilitating its spread to other grain crops.

Life History. Several species of cherry (*Prunus* spp.) are the overwintering hosts of the bird cherry-oat aphid. However, migrant aphids from southern winter wheat are more important than overwintering eggs in reestablishing populations in the north each spring. Bird cherry-oat aphids complete several generations on small grain crops before colonizing corn in July and August.

R. W. Kieckhefer

Corn Leaf Aphid

Scientific Classification. *Rhopalosiphum maidis*, formerly *Aphis maidis*, Homoptera (Aphididae).

Origin and Distribution. The corn leaf aphid is a worldwide pest of grasses between latitudes 40°N and 40°S. In North America, corn leaf aphids do not overwinter north of Texas; however, winged forms migrate each spring, extending their range into Canada.

Description. Populations of corn leaf aphids are comprised of winged and wingless females. Winged corn leaf aphids are about 1/16 in (1.6 mm) long with clear wings and black eyes, thorax, and legs. The abdomen is pale blue-green with black spots on each side and with lighter, transverse black bands. Two black cornicles are evident at the rear of the abdomen. The base of each cornicle is surrounded by a black spot. Adult wingless corn leaf aphids are blue-green and 1/16–3/32 in (1.6–2.5 mm) long. The cornicles, bases of cornicles, and legs are black. As they age, the adults become almost completely dark green to black. The four nymphal instars look very similar to wingless adults, except the nymphs are smaller.

Pest Status. The corn leaf aphid is an occasional pest of corn, sorghum, and barley and is also the primary vector of MDMV, a major disease of corn in the southern and eastern corn-growing areas of North America. This virus causes stunting and mottling of corn plants.

Injury. Nymphs and adults feed primarily in the whorls of corn plants, removing water and nutrients from the phloem. The aphids excrete honeydew from the anus. If corn plants have adequate moisture, corn leaf aphids cause little damage. However, feeding by aphids during drought conditions exaggerates symptoms of drought stress. The feeding and secretions by aphids may result in yellowing or reddening of the whorl leaves, sticky gummed-up leaves, sticky tassels that shed pollen poorly, tassels with dead sections, and barrenness or poor grain fill. Winged aphids feeding on newly emerged seedlings can reduce yield if the seedlings are not vigorous.

Corn leaf aphids (Marlin E. Rice).

Life History. Corn leaf aphids can complete as many as 40–50 generations per year. They reproduce without mating (parthenogenesis) and give birth to live nymphs. The aphids overwinter on Johnsongrass and other grasses in the southern part of their range. Winged adults migrate north in early spring and initiate colonies of wingless aphids on corn and other grasses. Colonies expand rapidly in moist corn whorls until tassel emergence, after which time the aphids move downward on the plants, forming small colonies behind husks and on stalks and leaves, surviving until the plants senesce. An abundance of winged aphids are produced during tassel emergence and senescence, and these forms migrate to more suitable hosts.

J. W. Bing

Corn Root Aphid

Scientific Classification. *Aphis maidiradicis*, formerly *Anuraphis maidiradicis* and *Aphis maidi-radicis*, Homoptera (Aphididae).

Origin and Distribution. The native corn root aphid is found throughout the United States east of the Rocky Mountains and is most prevalent in the Corn Belt.

Description. The wingless adult corn root aphid is a small, spherical, soft-bodied insect about 1/16 in (1.6 mm) long, and generally is blue-green to gray-green with a black head, thorax, and first abdominal segment. A white, waxy pubescence often makes them appear gray-green. The four nymphal instars are similar to the wingless adult but are

Corn root aphids (Marlin E. Rice).

Greenbugs (Marlin E. Rice).

smaller and lighter green. Winged forms have a black head and thorax, a green abdomen, and clear wings. The abdomen has black markings on each side of the second, third, and fourth segments, and transverse black marks on the last two segments. Black cornicles are located on the rear of the abdomen.

Pest Status. The corn root aphid is an economic pest only on corn, but it feeds on a number of weeds. Although the corn root aphid once was a serious pest of corn, economic damage by this insect has become less common.

Injury. Corn root aphids feed on corn roots and remove nutrients. Injury often results in dwarfing, yellowing, and reddening of young plants. External symptoms of injury are not apparent when feeding occurs on older plants.

Life History. The corn root aphid has both a sexual egg-laying form and another form in which only females are present and give birth to live nymphs. The sexual form lays eggs in the soil in the fall. Cornfield ants collect the eggs and nurture them through the winter. The eggs hatch in the spring, and the ants transfer the nymphs to roots of weeds and corn. During the summer, the sexual form is not present and females give birth to live nymphs. The ants continually tend the aphids, feeding on their honeydew. Winged aphids emerge from the soil during the summer, take flight, and with the help of ants, colonize other cornfields.

J. W. Bing

Greenbug

Scientific Classification. *Schizaphis graminum*, Homoptera (Aphididae). Also known as the spring grain aphid, spring grain aphis, bluegrass aphid, green louse, and Texas louse.

Origin and Distribution. Although greenbugs probably originated in southern Russia, they now are distributed worldwide. Greenbugs were discovered in the United States in 1882 and are found from Canada to the Gulf states and from the Atlantic to the Pacific. It is most important as a pest in Texas, Oklahoma, Colorado, Kansas, Nebraska, and South Dakota.

Description. Although greenbugs may be winged, most individuals in economically important infestations are wingless. The pear-shaped body, 1/16–3/32 in (1.6–2.1 mm) long, is yellow-green, although the abdomen may appear blue-green. A dark green stripe is evident along the middle of the back. The cornicles are pale with dark tips.

Pest Status. The greenbug occurs on more than 70 species of host plants, but it is economically important primarily on small grains and sorghum. Although found often on corn, it causes no economic damage. A number of distinct races or biotypes differ in their ability to infest different crops. Biotypes A through K have been described. Before the development of biotype C in 1968, the greenbug probably was no more important on sorghum than it currently is on corn. However, the greenbug became a severe pest of sorghum with the appearance of biotype C. Biotype E greenbugs feed on sorghum but can flourish on certain corn genotypes. Therefore, selection pressure could cause a shift to a biotype that is virulent on corn. Because the greenbug is injurious to host plants, a new biotype virulent to corn would elevate its pest status.

Injury. Greenbug colonies usually are located on the undersides of lower leaves. They injure their hosts by sucking phloem sap, but they inflict far more damage by injecting salivary toxins that induce a variety of responses by the plant. Most sorghum hybrids form distinctive red spots close to feeding sites; different types of symptoms occur on other host species. Although any plant stage can suffer damage, seedlings are especially susceptible. When aphid densities are large, small plants may die and more mature plants may lose many of their leaves. The greenbug also may transmit barley yellow dwarf virus in small grains and MDMV in sorghum. If the greenbug becomes a pest of corn, its importance in transmittng these viruses should be evaluated.

Life History. In the southern United States, greenbugs reproduce asexually all year and give birth to live nymphs. Without mating, a female can produce 2–6 offspring (all female) per day over a 25-day period. A newborn can mature and produce offspring by herself within 7 days of birth. Because of their reproductive potential, greenbug populations can increase in size rapidly, resulting in a damaging infestation at any time during the growing season. The first females of the spring may be winged and carried north by wind currents. North of 36° to 38° latitude, winged males and females may be produced in the fall, and, after mating, females may lay eggs that overwinter. However, this scenario has not been well documented. In the spring, eggs hatch into females that give birth to nymphs.

John C. Reese and Clinton Pilcher

Potato Aphid

Scientific Classification. *Macrosiphum euphorbiae*, formerly *M. solanifoli*, Homoptera (Aphididae).

Origin and Distribution. The potato aphid most likely originated in North America but now has a worldwide distribution.

Description. Potato aphids are medium sized, pear-shaped aphids that are 1/16–1/8 in (1.7–3.6 mm) long. They usually are green but can be yellow, pink, or magenta with distinct red eyes. Winged forms are green with yellow lobes on the thorax; their antennae and cornicles are darker than on the wingless forms.

Pest Status. Potato aphids are damaging to many vegetable crops, especially potatoes and tomatoes. Although these aphids are minor pests of corn, they can transmit MDMV.

Injury. Potato aphids generally form small colonies on the undersides of corn leaves. Symptoms of feeding injury appear infrequently on corn.

Life History. Eggs overwinter on wild and cultivated roses and hatch in the spring into wingless females that produce both winged and wingless offspring. The winged forms move from roses to other hosts such as corn. In the fall, winged individuals move back to roses where sexual forms mate and the females lay eggs that overwinter.

J. W. Bing

Management of Aphids

Natural enemies often regulate populations of aphids, so conservation of natural enemies can be an effective method of managing aphids in corn. Lady beetles, lacewings, and syrphid fly maggots prey on all species of aphids. The wasp *Lysiphlebus testaceipes* parasitizes corn leaf aphids and can destroy summer outbreaks of greenbugs. On a resistant plant variety, parasitoids may virtually eliminate greenbugs from a field before much damage occurs. A fungal disease also may suppress aphid populations. Heavy rainfall sometimes reduces the size of aphid colonies.

Several cultural tactics help manage aphids in corn. Crop rotation historically has controlled corn root aphids, and early plowing before planting keeps wild hosts under control and reduces densities of corn root aphids. Because greenbugs tend to settle more frequently into cleanly tilled fields, conservation tillage is recommended for their management. Destruction of volunteer grains, especially wheat, also may reduce greenbug densities. Controlling Johnsongrass reduces the density of the overwintering host of MDMV. Growing corn varieties resistant to corn leaf aphids reduces the size of colonies. Resistant sorghum hybrids help manage greenbugs. Crop hybrids resistant to diseases transmitted by aphids reduce the incidence of the diseases.

Insecticide applications may be considered for corn leaf aphids if colonies become large on drought-stressed corn. Although economic thresholds vary, moderate infestations (50–400 corn leaf aphids per plant) on 50% of the plants may warrant control.

Selected References. 22–24, 33, 51, 58, 68, 83, 90, 113, 114, 166, 181, 223, 238, 257

J. W. Bing, R. W. Kiekhefer, and John C. Reese

Armyworms

Three species of armyworms can cause significant injury to corn in North America. Because each species poses a different threat, they are discussed individually.

Armyworm (Marlin E. Rice).

Armyworm

Scientific Classification. *Pseudaletia unipuncta*, formerly *Cirphis unipuncta*, Lepidoptera (Noctuidae).

Origin and Distribution. The armyworm is native to North America and is distributed throughout the United States east of the Rocky Mountains and into southern Canada. It is an occasional pest of corn in the north-central and western Great Plains regions.

Description. The adult is pale gray-brown to pale brown with a 1-in (2.5-cm) wingspan. A white dot is evident in the center of each front wing. Small white eggs are deposited in rows or clusters on the surfaces of lower leaves. The full-grown larva is 1 1/2–1 3/5 in (3.7–4.1 cm) long and green-brown with varying degrees of black mottling and white flecks. Two orange stripes along each side and two dark stripes on the back are characteristic. The head is yellow-brown with a brown netlike pattern of dark lines. Armyworm larvae can be confused with fall armyworm larvae; however, the fall armyworm usually has a white inverted **Y** on the front of its head. The armyworm pupa is dark brown and can be found in an earthen cell below the soil surface.

Pest Status. The armyworm feeds primarily on wild and cultivated grasses, especially corn, timothy, millet, bluegrass, and small cereal grains, and on some legumes. Armyworms cause problems more frequently in reduced-tillage cornfields than in conventionally tilled cornfields. Armyworms also frequently invade corn grown near cereal grain fields. Females deposit eggs on the grasses in reduced-tillage or no-till fields before corn is planted. If a herbicide is used to control the grasses, larvae move from the dead grasses to the corn. If armyworm eggs are deposited in cereal grain and the larvae that hatch have not completed development when the grain matures, they migrate to corn to complete their development. Armyworm larvae migrate in large numbers and are capable of defoliating a cornfield in 1–2 days.

Injury. Armyworm larvae may feed only on leaf margins, or they may strip the plants, leaving only the stalks and leaf midribs. A corn plant recovers from this injury when feeding activity is moderate, as long as the growing point of the plant has not been injured. However, entire cornfields can be defoliated when an armyworm infestation is heavy and feeding damage is severe.

Life History. The armyworm completes 2–3 generations per year. The first generation occurs in May and June, and second and third generations occur in July and August, respectively. Armyworms overwinter as pupae in the soil in warmer climates. Adults migrate in April and May from southern parts of their range into northern corn-producing states. Eggs are deposited at night in rows or clusters on the lower leaves of grasses. Adults remain hidden during the day. Larvae feed at night and can be found during the day in leaf sheaths, ground litter, or soil. Pupation takes place in the soil or ground litter in a cavity or cell. Adults emerge in about 1 week, except for the generation that overwinters.

Management. Elimination of grassy weeds from cornfields or from field borders reduces the potential for armyworm infestations and outbreaks. If armyworm numbers are large and depletion of their grassy weed hosts causes migration into corn, an insecticide can be used for control. If the grassy weeds are patchy, treatment can be confined to infested areas. An insecticide application to seedling corn may be warranted when stand loss exceeds 10%. Armyworm out-

Armyworm injury to corn (Marlin E. Rice).

Armyworm injury to cornfield (Marlin E. Rice).

(*Left to right*) Heads of armyworm, fall armyworm, and corn earworm (Marlin E. Rice).

Fall armyworm (Marlin E. Rice).

breaks usually are suppressed in warm, dry weather by naturally occurring parasitoids and predators. The braconid wasp *Cotesia marginiventris* is one of the most common parasitoids of armyworms.

Fall Armyworm

Scientific Classification. *Spodoptera frugiperda*, formerly *Laphygma frugiperda*, Lepidoptera (Noctuidae).

Origin and Distribution. The fall armyworm is a tropical species and a continuous resident of the Gulf Coast states; the tropics of North, Central, and South America; and some of the West Indies. Adults migrate northward each year from southern regions when weather conditions permit. The fall armyworm is distributed throughout most of the United States east of the Rocky Mountains, from Maine to Florida in the east, and from Texas to Montana in the west.

Description. The adult is dark gray, light and dark mottled with a 1–1 3/4 in (2.5–4.2 cm) wingspan. A white spot is evident near the tip of the front wing. Eggs are spherical, gray, and covered with hairs and wing scales from the female. They are deposited in clusters on leaves of host plants. The mature fall armyworm larva is about 1 1/4 in (3 cm) long and varies from light tan or green to black. The larva is smooth skinned with three yellow-white lines along the back and a wider dark stripe on each side of the yellow-white lines. Below the dark stripe on each side is a wide, wavy yellow stripe with red splotches. The head is dark brown with a white inverted **Y** on the front. Fall armyworm larvae can be confused with corn earworm and armyworm larvae. However, a corn earworm larva has a yellow-brown head and rough skin covered with microspines; armyworm larvae do not have the white inverted **Y** on the head. The fall armyworm pupa is dark brown and 3/5–3/4 in (1.5–2 cm) long.

Pest Status. The fall armyworm feeds on many broadleaf

Fall armyworm larva (Marlin E. Rice).

plants and grasses, including corn, bermudagrass, and rice. Fall armyworms can injure corn from seedling emergence through maturity in the southern areas of the Corn Belt but are only occasional pests in the north-central and western Great Plains regions. It is an infrequent pest of many crops.

Injury. Early generations in the southern United States feed in the whorls of corn plants, causing defoliation. Corn plants recover when feeding activity is moderate, as long as the growing point of the plant has not been injured. Injury to the growing point results in serious damage or stunting of the plant. Late-planted, seedling-stage corn is more susceptible to larval feeding injury than early planted corn. Larvae feeding on corn at the silking stage can partially or completely destroy the tassel, ear, and leaves of the upper portion of the plant. Larvae also feed on the kernels as the ear develops.

Life History. The fall armyworm completes 1–3 generations per year in northern states and 5 or more generations in southern states. The life cycle from egg to adult takes 30–50 days, depending upon the temperature. Fall armyworms overwinter as partly grown larvae in southern states along the

Injury from fall armyworm (Marlin E. Rice).

Gulf Coast and cannot survive in areas where the ground freezes during the winter. In the spring, after fall armyworms pupate and emerge as adults, they migrate as weather conditions permit. Northern migration continues throughout the summer and into the fall. Egg masses are deposited at night on corn leaves and on other vegetation. Adults remain hidden during the day. Eggs hatch in 2–10 days, depending upon temperature. Larvae feed during daylight in the whorl or ears for 2–3 weeks. Pupation takes place in the soil, and the adult emerges after 10 days to 2 weeks.

Management. Control of fall armyworms with insecticides to prevent foliage-feeding injury usually is not economical, unless the infestation is severe and the larvae are present on areas of the plant where they will be exposed to the insecticide. An insecticide application may be warranted if egg masses are found on 5% or more of the plants, or if more than 50% of the plants exhibit severe leaf injury. Insecticidal control of larvae feeding in the ear usually is impossible. Parasitoids and predators provide some suppression of fall armyworm populations. The braconid wasp *Cotesia marginiventris* is one of the most frequently encountered parasitoids of fall armyworms. In the middle latitudes of the United States, avoidance of late planting and planting long-season hybrids usually prevents the coincidence of susceptible crop stages and large numbers of migrating adult fall armyworms.

Some *Bt* corn hybrids offer partial control of fall armyworms by suppressing their populations. These hybrids also reduce the amount of fall armyworm injury to the ear, thus decreasing the level of mycotoxins (e.g., aflatoxins) in the grain. Refer to the "European Corn Borer–Management" section (p. 81) for more details about transgenic corn and resistance management strategies.

Yellowstriped Armyworm

Scientific Classification. *Spodoptera ornithogalli*, formerly *Prodenia ornithogalli*, Lepidoptera (Noctuidae).

Origin and Distribution. The yellowstriped armyworm is a tropical and subtropical species that migrates northward as weather conditions permit. It is distributed throughout the eastern half of the United States to southern Canada, and also in Arizona, New Mexico, and southern California. The yellowstriped armyworm is most abundant in the southern United States.

Description. The adult is gray and dark mottled with a 1–1 1/2-in (2.5–3.8-cm) wingspan. The front wings are

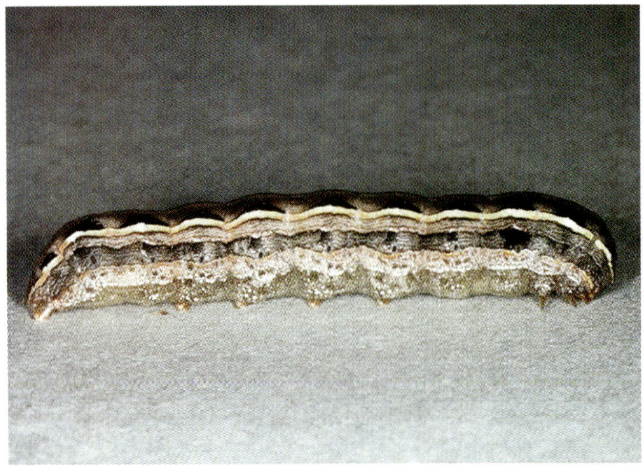

Yellowstriped armyworm (Marlin E. Rice).

light to dark gray with a slanting yellow-white bar near the middle of each wing. The mature yellowstriped armyworm larva is 1–1 2/5 in (2.5–3.5 cm) long, glossy black to gray with dark triangular spots along the midline of the back. A yellow stripe is evident along each side of the triangular spots, and there is a black spot above the hind legs. The head is brown with a netlike pattern. Larvae of the yellowstriped armyworm are similar in appearance to western yellowstriped armyworms, but yellowstriped armyworms have less extensive and less obvious netlike patterns on the head. In California, where the distributions of these two species overlap, the yellowstriped armyworm does not occur north of the Techachapi Mountains.

Pest Status. The yellowstriped armyworm feeds on grasses, including corn. It is a pest in the southern United States but seldom a problem in the north-central and Great Plains regions of the Corn Belt.

Injury. The larvae are occasional pests of corn, feeding on the leaves of young corn plants or in the whorl. Moderate infestations are not harmful, as long as the growing point of the plant is not injured.

Life History. Three or more generations occur each year in the southern states. Females deposit eggs in irregular masses covered with body scales on plant foliage. Larvae feed during the day. The pupa is the overwintering stage, and pupation takes place in the soil.

Management. The yellowstriped armyworm seldom is a serious pest of corn in the Corn Belt. Control with insecticides is not economical unless feeding would cause heavy damage (defoliation greater than 50%).

Selected References. 18, 48, 123, 173, 182, 234

Sharron S. Quisenberry

Billbugs

Scientific Classification. Southern corn billbug, *Sphenophorus callosus*; maize billbug, *S. maidis*; other *Sphenophorus* spp., Coleoptera (Curculionidae).

Origin and Distribution. Billbugs are native to North America. They are most common in the mid-to-lower Atlantic Coastal Plain but also are found throughout the southeastern United States, and in the Corn Belt, Texas, and Arizona.

Description. The adults are large, measuring 1/4–1/2 in (7–13 mm) long, and grey, dark brown, dull black (southern corn billbug), or black with shiny black ridges on the thorax and wing covers (maize billbug). Eggs are kidney shaped, white, and 1/10 in (2.3 mm) long. Larvae are white, legless grubs, widest at the posterior, and are 5/8 in (16 mm) long when full grown. The head is brown to mahogany. The pupa has all adult appendages visible and is white at first but turns yellow, then red.

Maize billbug (Marlin E. Rice).

Southern corn billbug (Marlin E. Rice).

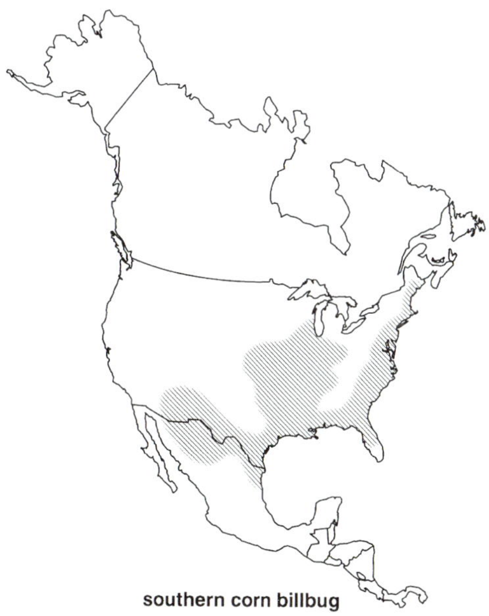

Pest Status. Southern corn and maize billbugs are infrequent, local pests in most corn- growing areas but are common pests in the mid-to-lower Atlantic Coastal Plain. Billbugs can be the most important insect pest in this region, especially on poorly drained or organic soils. Infestations occur most frequently in nonrotated corn, along field edges, and in areas populated with nutsedge, an alternate host. The potential for billbug problems is greater in no-till corn than in fields that have been tilled.

Injury. Adults injure corn seedlings up to the 6-leaf stage by feeding in the lower stem. Feeding slits occur on the lower stem, and the whorl leaf often is cut inside the plant. Small plants die or tiller extensively if the growing point is injured; otherwise, rows of oblong holes in expanded leaves or leaves that have been cut off are the only results of adult billbug feeding. As stems become larger in diameter, feeding by adults has little effect. If billbugs are numerous when seedlings emerge, damage can be sudden and severe. Damage usually is most severe along field edges but sometimes is spread throughout the field.

Billbug larvae can reduce corn yield by as much as 50% by eating a cavity into the root crown and by feeding in the lower stalk.

Life History. Billbugs complete 1 generation per year. The adult stage overwinters, primarily in protected sites along field edges. Emergence of the adults in the spring usually coincides with the time of corn planting and lasts for 5 or more weeks. Adults move primarily by walking, although they can fly if they are stressed. After feeding on corn or sedges, adults mate and the females lay eggs either in the soil or in cavities eaten into plant stems. After hatching, the larvae tunnel into the root crown or into the soil where they feed in the root mass. Larvae pass through 5 or 6 instars, pupate within a chamber, and emerge in midsummer to early fall. *Patasson calendrae*, a small wasp that parasitizes the egg, attacks several *Sphenophorus* species, including *S. callosus* and *S. maidis*.

Seedling corn injured by billbugs (Marlin E. Rice).

Management. Where billbugs are perennial pests, an integrated program of cultural and chemical tactics prevents current-season damage and population increase in the following season. Crop rotation is effective because billbugs do not move far and have a narrow host range. Crop rotation practices should emphasize isolation and consolidation of fields. Controlling sedges helps eliminate alternate hosts. Early planting and rapid growth reduce the period of crop susceptibility before all billbugs emerge and disperse into cornfields. In-furrow subsoiling also enables corn seedlings to tolerate injury. Proper fertility and pH maintenance, rapid-growing cultivars, tillage, and application of starter fertilizers help to reduce the period of crop susceptibility. Soil insecticides applied at planting kill adults and larvae.

Growers rely on scouting and treating with postemergence insecticides when other tactics fail to control billbugs. Postemergence insecticides are warranted when 5% of the seedlings are lost. Spot treatment along field margins reduces the amount of insecticide applied.

Selected References. 139, 171, 248

John W. Van Duyn and Robert J. Wright

Carrot Beetle

Carrot beetle (Thomas J. Riley).

Scientific Classification. *Ligyrus gibbosus*, Coleoptera (Scarabaeidae).

Origin and Distribution. The carrot beetle is native to and occurs throughout the United States, except in some of the far northern states, and into Central America.

Description. Adults are red-brown, oblong-shaped "June" beetles about 5/8 in (13 mm) long. The larvae are white to blue-white, fleshy grubs with a brown head and three pairs of well developed brown legs. The larvae usually are curved into a C shape. Depending upon the stage of development, the larvae may be 1/4–1 1/8 in (5–31 mm) long. Pupae are 5/8 in (15 mm) long and pale brown.

Pest Status. The carrot beetle is a sporadic pest of corn. Adults attack seedlings and larvae feed on roots throughout the season. The potential for infestations by carrot beetles is greater in corn planted after grass sod and in no-till corn than in conventionally tilled cornfields in other crop rotation schemes.

Injury. Adults feed on seedlings slightly below the soil surface, producing a large gougelike wound that damages or destroys the growing point and disrupts water conducting tissues. Plants are killed outright or growth is impaired so that injured plants are not productive. Adults that have overwintered nearby infest cornfields within 45 days after planting. Larvae feed on the roots later in the season but usually do not cause significant injury.

Life History. The carrot beetle completes 1 generation per year. Adults overwinter in deep soil and become active when temperatures warm in the spring, coinciding with planting and early growth of corn. Severe injury sometimes resembles injury caused by sugarcane beetles. Infestations often occur where corn is planted into grassy areas, especially in minimum tillage or no-till systems. The adults mate shortly after becoming active in the spring, and females deposit eggs at the bases of plants or in the soil from May through July. Eggs hatch in about 11 days; the larvae feed initially on decaying plant debris and then on the roots of a variety of plants, including corn. Larvae feed and develop for about 52 days, and pupation in the soil lasts about 1 week. New adults usually emerge during September and October and are active for a short time before entering diapause.

Management. Avoidance of planting corn into grass sod or other grassy areas that have been established for 2 or more years is a preventive control tactic for carrot beetles. No-till or minimum tillage practices increase the potential for infestation. Early planting and cultural practices that foster rapid and vigorous growth of seedlings enable plants to escape or to tolerate carrot-beetle-feeding injury in the spring. Some insecticides can be used for preventive control, but use of chemicals as a rescue treatment is questionable because reliable scouting methods for carrot beetles are lacking.

Selected Reference. 115

John N. All

Cattail Caterpillar

Scientific Classification. *Simyra henrici*, formerly *Leucania henrici*, Lepidoptera (Noctuidae).

Origin and Distribution. The cattail caterpillar is native to and widely distributed throughout North America.

Cattail caterpillar (Marlin E. Rice).

Description. The larva grows to 1 3/4 in (45 mm) long and is marked with yellow, orange, and black. Tufts of bristly, black-and-white hairs arise from orange, raised, wartlike areas on the top of its back. The back is black; the sides and underside are mottled black. Areas along each side are yellow. The head capsule is black with a white area on the "face" that forms a distinct, inverted **V**. The light brown, medium-sized adults resemble armyworm adults but are lighter and lack the white spot on each forewing. Two dark stripes along the middle of the forewings usually are evident.

Pest Status. The cattail caterpillar rarely attains pest status on corn; however, it has been reported to injure wheat. Other hosts are cattail, rye, canary grass, other grasses and sedges, smartweed, soybeans, and other broadleaves.

Injury. Small larvae eat only partway through a leaf, leaving a few layers of cells that are translucent. Such injury often is called "windowpaning." Injury caused by older larvae is similar to injury caused by grasshoppers; the caterpillars may consume all of the leaf except the midrib. Injury usually is limited to a few scattered plants, but occasionally plants in areas as large as an acre have been stripped of leaves.

Life History. This species completes 2–3 generations each year in the Corn Belt. Adults emerge from overwintering pupae about mid-April. A second flight period starts about mid-June, and a third flight peaks in mid- to late August. Eggs normally hatch in 5–7 days. The larval phase lasts 4–6 weeks, with 7–8 instars.

Management. Cattail caterpillars rarely are pests because numerous parasitoids, predators, and pathogens kill many of the larvae. Detection of populations that could be damaging when natural control fails is accomplished through routine scouting.

Selected References. 62, 63

Noel N. Troxclair, Jr.

Cereal Leaf Beetle

Cereal leaf beetle (Marlin E. Rice).

Scientific Classification. *Oulema melanoplus*, Coleoptera (Chrysomelidae).

Origin and Distribution. The cereal leaf beetle is originally from Eurasia. It is found in North America from Maine to South Carolina, west to Arkansas and Minnesota, and in Montana and Utah.

Description. Adults are 1/6–1/5 in (4–5 mm) long with shiny blue elytra, black body and head, and red pronotum. Their legs are orange to red with black tarsi. The larvae are 1/6–1/4 in (4–6 mm) long. They cover their thorax and abdomen with a mixture of mucous and fecal material that appears shiny brown to black. The larva is pale yellow to orange beneath the coating of fecal material.

Pest Status. The cereal leaf beetle is infrequently a pest of corn when small grains are not nearby. However, after completing their spring development in small grains, newly emerged adults often move to corn to feed before entering diapause.

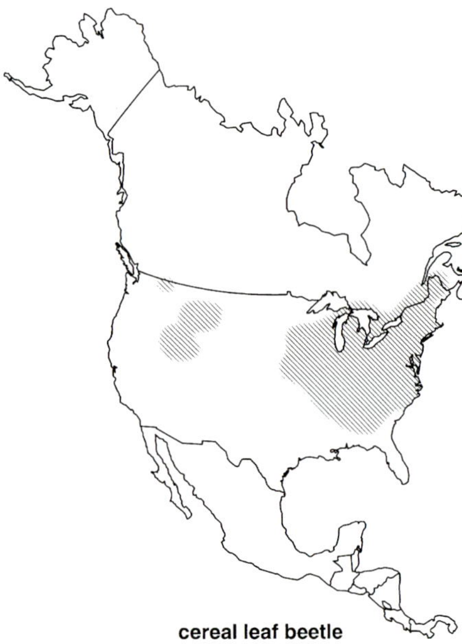

Injury. "Summer" adults feed on the upper surface of a leaf between the veins, often leaving the bottom epidermis intact. Economically damaging defoliation is rare.

Life History. In early spring, overwintering adults become active and feed for a short time before mating. Females glue oblong, yellow eggs to the surface of leaves; the eggs darken just before they hatch. Larvae feed for about 2 weeks. Pupation takes place in the soil over a period of 2–3 weeks. When soil moisture is high, pupation may occur on the plant. Emerged adults feed for about 3 weeks, often on corn, before entering diapause in sheltered areas. The cereal leaf beetle completes 1 generation per year.

Natural enemies of cereal leaf beetle eggs include the wasp parasitoid *Anaphes flavipes* and the lady beetle *Coleomegilla maculata lengi*. Larval parasitoids are *Tetrastichus julis*, *Diaparsis carinifer*, and *Lemophagus curtus*. The tachinid fly *Hyalomyodes triangulifer* attacks the adult.

Management. Control of adult cereal leaf beetles in corn is not necessary unless the crop is under stress, defoliation is greater than 50%, and large numbers are present.

Selected References. 169, 256

Michael J. Haas and Douglas A. Landis

Chinch Bug

Chinch bugs (Marlin E. Rice).

Scientific Classification. *Blissus leucopterus leucopterus*, Heteroptera (Lygaeidae).

Origin and Distribution. The chinch bug is native to North America and occurs throughout the United States in native grassland areas. The hairy chinch bug, *Blissus leucopterus hirtus*, and the southern chinch bug, *Blissus insularis*, occur in the northeastern and Gulf Coast states, respectively, but they rarely are pests in corn.

Description. Adults are small, elongate bugs about 3/16 in (5 mm) long and 1/16 in (1.5 mm) wide. They are mostly black with a white triangle in the middle of each wing making an irregular white band across the midsection of the body. Some adult chinch bugs have short wings that do not cover the rear of the body. Nymphs have no wings and are red for the first stages of their development. They turn black with a white band across mid-body for the last three stages of development, and small wing pads appear. All stages of this insect can occur in corn simultaneously.

Pest Status. The chinch bug is a sporadic pest throughout its range but can be a major pest when late planting is followed by dry conditions. Adult chinch bugs may migrate into corn in early spring. Nymphs and adults also infest corn after wheat or after other early spring host plants begin to mature and dry up. Hundreds of chinch bugs can infest a single plant. Seedling injury is most likely when dry weather retards plant growth or when the activity of insecticides applied with the seed at planting is reduced.

Injury. Injury caused by chinch bugs may be slight to severe, depending upon the conditions of infestation. Also, symptoms do not appear until several days after chinch bugs first infest plants. Chinch bugs cause damage by sucking sap from the plant base, mostly from behind leaf sheaths, but occasionally from roots. Feeding punctures cause physical injury, and sap removal deprives the plant of nutrients. Young nymphs cause little injury; older nymphs and adults cause the most damage. Symptoms of feeding injury include stunting (failure of leaf sheaths to elongate properly), red feeding marks behind leaf sheaths, yellow streaking on leaves, and a tendency for plants to curve as they grow. Severely damaged plants may die or be stunted, with leaves at the growing point wilted or dead.

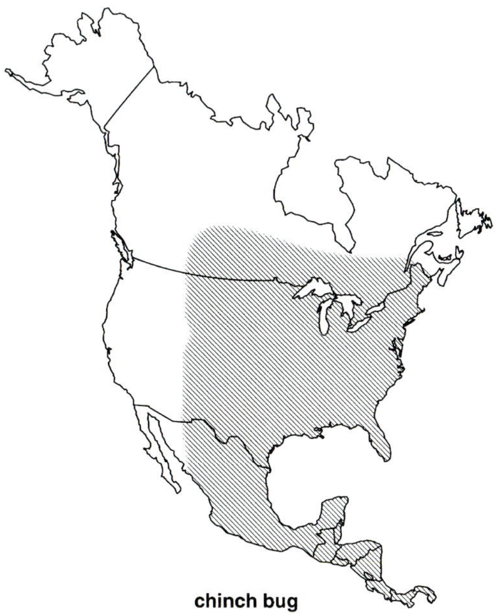

chinch bug

Chinch bug nymphs on corn (Marlin E. Rice).

Life History. Chinch bugs complete 3 or more generations per year in southern areas, and 2 in northern states. Adults spend the winter in prairie bunch grasses, primarily of the genus *Andropogon*. In late March and early April in southern areas and in April and May in northern areas, adults fly to host plants for feeding, mating, and oviposition. In southern areas, the adults infest seedling corn. Feeding, mating, oviposition, and the growth of nymphs occur in corn. Chinch bugs spend 2–3 generations in corn. Nymphs from the last generation can complete development on corn, sorghum, or wild grasses, then migrate into bunch grasses to overwinter. In some areas, chinch bugs first infest wheat, barley, or other small grains, where they feed, mate, lay eggs, and develop as nymphs until the crop matures. Nymphs then migrate by crawling as far as 1/4 mi (0.4 km) into corn.

Fungi in the genus *Beauveria* may infect overwintering chinch bugs during mild, wet winters, but the impact on overwintering populations is unknown.

Management. Early planting to avoid infestation by chinch bugs is highly recommended. Early planted corn usually is able to tolerate all but very heavy infestations of chinch bugs. Drought stress increases the potential for chinch bug problems. Adequate moisture enables plants to outgrow a chinch bug attack and ensures activation of granular soil insecticides applied at planting. Corn should not be planted next to wheat or barley in areas where damage occurs because nymphs migrate from small grains.

Several soil insecticides used to control corn rootworms and other soil-inhabiting pests also are effective against chinch bugs. In areas where chinch bugs pose a threat, a planting-time application of an insecticide registered for control of chinch bugs is warranted.

When other management tactics fail and damage caused by chinch bugs occurs, insecticide sprays can be applied. These sprays should be applied only with ground-operated equipment. A high-pressure spray in 20–40 gallons (84.6–169 liters) of water should be directed at the bases of the plants, ensuring that the insecticide contacts chinch bugs behind leaf sheaths and in the soil around the plant base.

When scouting for chinch bugs, one should check several areas because chinch bugs are not distributed evenly throughout infested fields, particularly when nymphs are migrating from nearby fields. Also, several plants should be examined because the number of bugs per plant varies.

When deciding whether to treat a chinch bug infestation, a grower should consider environmental conditions in the field and the growth stage of the plants. Young plants usually are more susceptible to injury than older plants, and drought conditions increase the potential for economic damage. Insecticide sprays are recommended when conditions are dry, injury is evident, and 5 or more chinch bugs are present on 20% of seedlings less than 6 in (15 cm) tall.

Selected References. 173, 179, 227, 231

Thomas J. Riley and Gerald E. Wilde

Corn Blotch Leafminer

Scientific Classification. *Agromyza parvicornis*, Diptera (Agromyzidae).

Origin and Distribution. The corn blotch leafminer is a native insect distributed throughout the continental United States and southern Canada.

Corn blotch leafminer adult and mine created by larva (Marlin E. Rice).

Description. The adult is gray to brown and about 1/4 in (6 mm) long. The egg is small, white, and oblong and is slightly constricted in the center. The larva is pale green to yellow-white and about 1/4 in (6 mm) long when mature. The puparium is red-brown and about 1/8 in (3 mm) long.

Pest Status. The injury caused by this insect is of little or no economic importance.

Injury. On small leaves, larvae begin feeding at the tip and move toward the base, consuming all green tissue between the upper and lower leaf surfaces. On larger leaves, injury may start anywhere; however, usually only the bottom half of the leaf is mined and the mine may not be visible from above. The tunnel widens as the larva grows, resulting in a broad blotch in the last third to half of the mine. Larvae seldom consume a large portion of a leaf, and usually only a few leaves per plant are injured.

Life History. Adults emerge in the spring and within 5–10 days, females begin to insert eggs into the leaves of corn. Females may lay 100 or more eggs. As larvae hatch, they tunnel in the leaf tissue, leaving transparent galleries or mines. At least 18 species of chalcid and braconid wasps parasitize the larvae. *Derostenus diastatae* probably is the most important parasitoid. Larvae drop to the soil to pupate, and adults emerge after 2–3 weeks. The life cycle is completed in 4–6 weeks. Corn blotch leafminers complete 3–5 generations each year in the north-central United States, fewer in the north, and more in the south. Leafminers overwinter as mature larvae in areas that do not freeze and as pupae farther north.

Management. No specific management practices are recommended.

Selected References. 188, 230

Robert J. Wright

Corn Delphacid

Scientific Classification. *Peregrinus maidis*, formerly *Delphax maidis*, *Pundaluoya simplicia*, Homoptera (Delphacidae).

Origin and Distribution. The corn delphacid originated in Africa on sorghum. It has spread to all tropical regions of the world, including oceanic islands where corn and sorghum are grown. In the United States, the corn delphacid is a sporadic pest of corn in the Gulf Coast states but a serious pest in Hawaii. The planthopper is highly sensitive to frost; thus, migrants that establish summer populations north of the Gulf Coast region are killed by winter weather.

Description. Short-winged (flightless) and long-winged (flying) adults occur. When viewed from above, the 1/10–

Corn delphacids (L. R. Nault).

1/8 in (2.5–3.5 mm) long, short-winged adults are pear shaped. The body is mottled with tan and brown markings. Stubby antennae arise from beneath the eyes. Each antenna bears a fine, hairlike terminal segment. The short wings do not cover the last four to five abdominal segments. The underside of the abdominal segments is black bordered in white. The long-winged adults are similar in appearance to the flightless adults, except they are larger (3/16–1/4 in [4–5 mm] long) and the wings extend well beyond the tip of the abdomen. The outer margin of the first pair of wings has tan or brown spots.

corn delphacid

Pest Status. The corn delphacid is a pest of both sorghum and corn throughout the insect's range.

Injury. Removal of plant sap by large numbers of corn delphacids may cause stunting of plants. Large amounts of excreted honeydew collect on plants and support the growth of sooty molds. More importantly, corn delphacids transmit plant viruses. Two viruses are transmitted exclusively by the corn delphacid, the maize mosaic rhabdovirus (MMV) and the maize stripe tenuivirus (MStV). The viruses occur in corn nearly everywhere the corn delphacid is found. Young plants inoculated with MMV develop fine yellow or white stripes and become severely stunted. Plants infected with MStV first reveal fine stripes that later coalesce into broad yellow or white stripes on upper leaves. The tips of some corn plants bend sideways. In susceptible corn varieties, early infection by either virus results in substantial yield losses.

Life History. Females insert eggs into leaf tissue with their sword-shaped ovipositors. At sea level in tropical climates, development from egg to adult takes about 4 weeks. Colonies of nymphs and adults frequently are found in the corn whorls. On older plants, dense colonies of mixed-aged nymphs and adults are found on the lower surfaces of leaves. The delphacid frequently is associated with ants that feed on honeydew and protect it from parasitoids and predators. When population densities are large, long-winged adults are produced, and they disperse to other corn or sorghum fields.

Management. Insecticides reduce delphacid populations and the spread of the viruses. Continuous, year-round production of corn promotes large densities of delphacids and encourages the spread of the viruses. A corn-free period is recommended to reduce numbers of delphacids and the incidence of the diseases. Varieties resistant to MMV have been developed.

Selected References. 102, 178, 224

Lowell R. Nault

Corn Earworm

Scientific Classification. *Helicoverpa (Heliothis) zea*, formerly *Heliothis zea*, Lepidoptera (Noctuidae).

Origin and Distribution. The corn earworm is native to the Americas and occurs wherever corn is grown in North and South America.

Description. Eggs are 0.5–0.6 mm in diameter. They are white to light yellow when first deposited, turning a dusky brown before hatching. The eggs are spherical with a flat base and numerous prominent ribs. Females prefer to lay eggs on newly emerged silks, but eggs also may be deposited on upper and lower leaf surfaces, and on stalks, tassels, and husks.

Early instar corn earworm (Marlin E. Rice).

Late instar corn earworm (Marlin E. Rice).

Larvae develop through five, sometimes six instars, reaching 1 5/8 in (40 mm) when full grown. Newly hatched larvae are translucent cream to white with a black head. Larger larvae vary from yellow, brown, and red, to green with prominent bands of cream, pink, green, or yellow. The head usually is dark yellow or orange. The cuticle is covered with microspines. Pupae are red-brown. Adult females are yellow-brown with brown forewings and a wingspan of about 1 1/2 in (3.8 cm). Adult males are light yellow with olive forewings. A dark spot is evident near the center of each forewing. Hind wings are light colored with a broad, dark brown outer band and usually a narrow, brown inter-marginal band.

Pest Status. The corn earworm is an important economic pest of corn, including sweet corn, popcorn, dent, flint, and floury corn. Larvae consume leaves, tassels, silk, and developing kernels. The most severe infestations occur in the southern United States where multiple generations occur. The large corn earworm populations that occur from June through September in the southern and southeastern United States originate from susceptible corn hybrids. Entomologists have estimated that densities of adult female corn earworms emerging from 494,000 acres (200,000 hectares) of corn are as large as 74–358 million moths, capable of producing economic infestations on 7.5–35 million acres (3–14 million hectares) of other crops. Losses caused by corn earworms in field corn have been estimated at 2.5% annually, although losses in the southeastern United States may range from 1.5–16.7%. Losses attributable to corn earworms in sweet corn can be as high as 50%.

Injury. Injury is caused by the larvae. Small larvae produce pinhole feeding injury on the leaves and eat tender tassels and silks. Larger larvae may feed on leaves, causing severe defoliation; severely damage the tassel, which can result in poor pollination; and consume the silks and developing kernels. Both young and old larvae feed in the silks; those that penetrate the ear destroy many of the developing kernels, particularly in sweet corn. Injury to the ears permits the invasion of secondary pests and microorganisms that may produce mycotoxins, resulting in additional losses.

Life History. Continuous generations occur in the extreme southern United States, especially in mild winters. Elsewhere, the corn earworm overwinters as a pupa in a protective cell 2–4 in (5–10 cm) deep in the soil. Adults emerge in March and April near latitude 31°N. Sweet corn, tomato, crimson clover, and geranium are important early hosts in the southern United States. Larval development in whorl-stage corn requires 20–21 days. Full- grown larvae leave the plant and drop to the soil. Larvae that have infested the ear may bore through the side of the husk to leave the plant. Larvae penetrate the soil to a depth of 2–4 in (5–10 cm) where they form pupal cells and remain for about 14 days. First-generation adults emerge during May and are attracted to tasseling and silking corn. Multiple (6–10) overlapping generations occur thereafter.

In the northern United States, only 1 or 2 generations occur. Generally, only 2 generations of the corn earworm occur on corn, one infesting the whorl stage and the second infesting the ear stage. Adults are strong flyers that migrate from

Corn earworm damage to corn ear (Marlin E. Rice).

crops in northern Mexico, south Texas, south Florida, and North Carolina to central and northern locations in the United States and Canada. Adults become active about 1 hour after sundown and search for plant nectaries and other sources of sugar. On the day of emergence, adults begin to mate about 4 hours after sundown; oviposition begins on the second or third night. A female may deposit 1,000–2,000 eggs during 6–8 nights.

Management. Although the corn earworm can be one of the most damaging insects of field corn, injury is so evenly distributed and control is so difficult that farmers generally accept the loss. In the southeastern United States, control of corn earworm larvae may be warranted if 50% of the whorls are infested, but control is not recommended to prevent larvae from entering the ears of field corn. Management of the corn earworm in field corn has been limited to growing resistant hybrids and planting early to avoid damaging densities of this pest.

Several natural enemies and pathogens limit the amount of damage caused by corn earworms. Predators and parasitoids of eggs and larvae include *Orius insidiosus*, *Apanteles marginiventris*, *Cardiochiles nigriceps*, *Microplitis croceipes*, *Campoletis sonorensis*, *Trichogramma* spp., and *Archytas marmoratus*. *Diapetimorpha introita* is a pupal parasitoid. Nematodes that affect prepupae and pupae are *Steinernema carpocapsae* and *S. riobravis*. Pathogens that infect the larvae include nuclear polyhedrosis virus and *Bacillus thuringiensis*.

Chemical control of earworms in corn is prohibitively expensive except to protect sweet corn, where much of the market value is in the quality of intact ears. Larvae feeding on the kernels cause the most important damage. Larvae usually feed on the silks and then move into the ears, leaving their waste and creating a highly favorable microclimate for other pests and especially for the production of mycotoxins. If the injury is slight, sweet corn growers can clip the tips of the ears and sell them at a reduced price; however, some markets accept only nondamaged and nonclipped ears.

Sweet corn plants should be examined weekly before silks appear. The first insecticide application is warranted only after late-whorl leaves or tassels appear damaged. Thereafter, insecticides for control of corn earworms should be applied weekly or more often. For control of larvae in the ear, treatments are warranted when silks first appear. When infestations are heavy, treatments are continued every 24–48 hours until all silks are dry or brown. In Florida, sweet corn grown between October and March is treated every 48 hours during the silking period. From April through September, controls are applied every 24 hours during the critical period because of the extremely large numbers of corn earworms present.

The present grading system for sweet corn often penalizes ears that have minor injury caused by larvae, so growers follow an intensive spray schedule to be able to market their product. Treatment of sweet corn from south Florida to south Georgia requires from 14–28 applications of insecticide per crop.

Resistant corn limits the amount of both leaf and ear injury caused by corn earworms. The most effective type of resistance in ear-stage corn is a combination of silks that are antibiotic to the larvae and ear coverage by a husk sufficiently tight to alter larval behavior. Larvae then initiate feeding in the ear tip, resulting in the production of small larvae or larvae that leave the ear before they complete development.

Some *Bt* corn hybrids offer partial control of corn earworms by suppressing their populations. These hybrids also reduce the amount of injury to the ear, thus decreasing the level of mycotoxins (e.g., aflatoxins) in the grain. Refer to the "European Corn Borer–Management" section (p. 81) for more details about transgenic corn and resistance management strategies.

Selected References. 68, 182, 195, 229, 259, 260

B. R. Wiseman

Corn Rootworms

Four kinds of corn rootworm attack corn in North America. Because each poses a different threat, they are discussed individually. The "Injury" and "Management" sections apply to all four corn rootworms.

Mexican Corn Rootworm

Scientific Classification. *Diabrotica virgifera zeae*, Coleoptera (Chrysomelidae).

Mexican corn rootworms (J. W. Stewart).

Mexican corn rootworm

Origin and Distribution. The Mexican corn rootworm is a native of Central America and occurs from Costa Rica north through Guatemala, eastern Mexico, central Texas, and southern Oklahoma.

Description. Mature larvae are 3/8 in (9–10 mm) long. Larvae occur in the soil where they feed within the roots of corn. Mexican corn rootworm larvae are similar in appearance to western corn rootworm larvae. The adult looks similar to the northern corn rootworm but is easily distinguished by a black stripe on the femur. The wing covers of the Mexican corn rootworm are light green without black stripes.

Pest Status. The Mexican corn rootworm is a serious pest of corn not rotated with other crops. Severe injury to brace roots causes lodging or plant death.

Life History. The Mexican corn rootworm completes 1 generation per year. In south Texas, eggs hatch in early April. Larvae chew into the brace roots of corn to feed. The larval stage is completed in 30–40 days. Larvae leave the brace roots and pupate in an earthen cell. Beetle emergence occurs over 25–30 days, during mid-May to mid-June in southern Texas. After feeding and mating, females lay eggs in soil cracks near the bases of corn stalks. Eggs remain dormant in the soil until the following April.

Adults emerge in northern Texas 1 month later than in southern Texas and approximately 3 months later in central Mexico. The later emergence in central Mexico is an adaptation to the rainy season.

J. W. Stewart

Northern Corn Rootworm

Scientific Classification. *Diabrotica barberi*, formerly *Galleruca longicornis*, *D. longicornis*, *D. longicornis barberi*,

Northern corn rootworm (Marlin E. Rice).

Coleoptera (Chrysomelidae).

Origin and Distribution. The northern corn rootworm is a native of North America, first reported in Colorado. Until 1955, it was distributed extensively but primarily throughout the north-central Corn Belt. In the late 1940s, however, the practice of planting corn after corn facilitated the pest's spread. Now it is found from the Rocky Mountains to the east coast and from southern Ontario to Tennessee.

Description. The larva is white with a brown head and a brown plate on top of the last abdominal segment. It is very small, 1/16 in (1.6 mm) long, just after egg hatch, but grows to 5/8 in (16 mm) at maturity. It has three pairs of short legs. The adult is 1/4 in (7 mm) long, oval, and pale to dark green, without markings.

Pest Status. Northern corn rootworm larvae routinely occur in cornfields in which corn was planted the previous season and are found frequently in the same fields with western corn rootworms. These two species are the most eco-

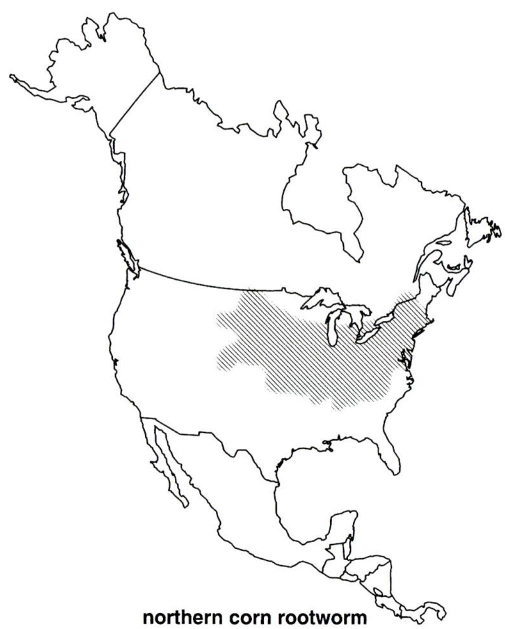
northern corn rootworm

nomically important insect pests of corn throughout the Corn Belt. Together they account for nearly $1 billion annually in crop losses and control costs.

Life History. Although a small percentage of some populations of northern corn rootworms takes longer than 1 year to complete development (due to extended diapause; see "Management of Corn Rootworms, Crop Rotation), most complete 1 generation per year. They overwinter as eggs in the soil. Most of the eggs are distributed throughout the top 8 in (20 cm) of soil, but some may be as deep as 12 in (30 cm) in dry soil. Eggs begin hatching in late May or early June. The larvae feed almost exclusively on corn roots and require about 3 weeks to complete development, depending on soil temperature. Pupation occurs in cells constructed in the soil and lasts for about 2 weeks.

Adult emergence from the soil begins in early July and continues throughout the summer. Males emerge first, followed shortly by females. Depending on the season, peak emergence occurs around late July or early August. Although adults feed on corn leaves, pollen, and soft kernels, silks are their preferred food. The adults mate shortly after emerging, and females develop mature eggs within 2 weeks after emergence. Females crawl into cracks in the soil and lay groups of eggs throughout the remainder of their lives. They deposit most of their eggs throughout the latter half of August and into early September. Although the beetles feed on the pollen of other flowers, they return to corn to lay eggs.

Jon J. Tollefson and Eli Levine

Southern Corn Rootworm (Spotted Cucumber Beetle)

Scientific Classification. *Diabrotica undecimpunctata howardi*, Coleoptera (Chrysomelidae). The larva is called the southern corn rootworm; the adult is called the spotted cucumber beetle.

Origin and Distribution. The southern corn rootworm is a widely distributed native insect commonly found east of the Rocky Mountains from southern Canada to northern Mexico. It also has been reported from Arizona, California, New Mexico, and Utah. However, damage to corn caused by southern corn rootworms is most common in the coastal regions of Texas and Louisiana.

Description. The adult is about 1/4 in (6 mm) long and is yellow-green with a black head and antennae and 12 black spots on the wing covers (elytra). The larva is cream colored, about 3/4 in (19 mm) long when full grown, and has six short legs on the thorax behind the brown head, with characteristic projections on the anal plate on the last abdominal segment.

Spotted cucumber beetle (Marlin E. Rice).

Pest Status. The southern corn rootworm is a minor pest of corn except in areas of Texas and Louisiana where damage can be widespread and severe, requiring insecticide treatment to protect corn stands early in the production season.

Life History. In contrast to other corn rootworms that overwinter as eggs, the southern corn rootworm overwinters as an adult in the southern United States. The adults become active when temperatures exceed 70°F (21°C) and feed on many types of plants, including weeds and grasses. They enter cornfields soon after plant emergence and deposit eggs. Eggs do not diapause; they hatch in 5–11 days at 72–86°F (22–30°C). Young larvae crawl through the soil and are attracted to corn roots. Larval development through three instars occurs in 16–9.5 days at 70–88°F (21–31°C). The pupal stage lasts for 12–5 days at 70–93°F (21–34°C). The length of one life cycle is 19.5–39 days. One or more generations develop on field corn per season.

Bastiaan M. Drees

southern corn rootworm

Western Corn Rootworm

Scientific Classification. *Diabrotica virgifera virgifera*, formerly *D. virgifera*, Coleoptera (Chrysomelidae).

Origin and Distribution. The western corn rootworm is widely distributed throughout the major corn-growing regions of the United States and in southeastern Canada, except west of the Rocky Mountains and in the southeastern United States. The species originated in the tropics or subtropics and occurs only in the New World.

The western corn rootworm was first recognized as a pest of corn in Colorado and western Kansas in 1912. Before the 1950s, this species occurred only in states along the eastern slope of the Rocky Mountains from northern New Mexico to Nebraska. It was not a major pest until growers began planting continuous corn in western Nebraska and Kansas during the mid-1900s. Its distribution extended eastward at a rate of 40–50 mi (64–80 km) per year until reaching Delaware and Maryland in the mid-1980s.

Description. Three of the four life stages of western corn rootworms reside in the soil. Eggs are oval, creamy white, and measure about 0.3 by 0.5 mm. Larvae have a dark brown head capsule and anal plate and creamy white thoracic and abdominal segments. Fully developed third instars are about 1/2 in (12 mm) long. Pupae are found in earthen cells in the soil and also are creamy white. Adults vary from light yellow to light green and have three black stripes on their wing covers. However, the black markings vary greatly; some adults have minimal markings and may be mistaken for related species such as northern or Mexican corn rootworms. Another pest often confused with the western corn rootworm is the striped cucumber beetle, *Acalymma vittata*; however, the underside of the striped cucumber beetle's abdomen is nearly black.

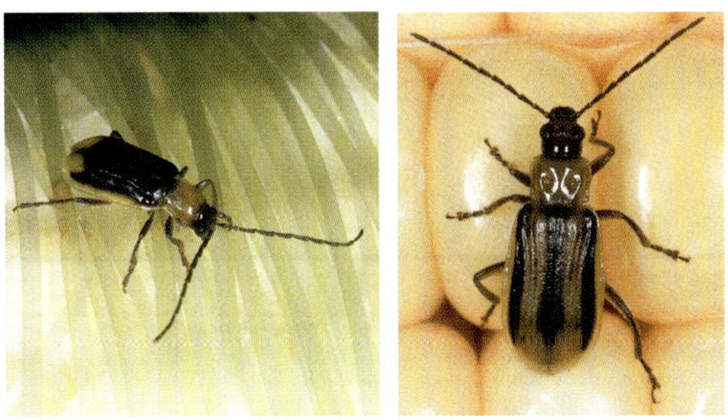

Male and female western corn rootworms (Marlin E. Rice).

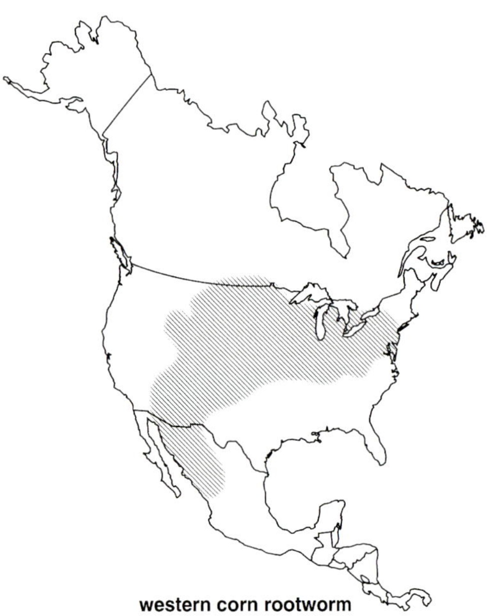

western corn rootworm

Pest Status. The western corn rootworm is one of the most serious pests of corn in North America. Corn is the only known crop host of the western corn rootworm. Females lay their eggs almost exclusively in cornfields; eggs hatch the following year. Until recently, nearly all crop damage could be avoided with annual crop rotation. However, in recent years in east-central Illinois and northwestern Indiana, western corn rootworm larvae have caused severe damage to corn planted after soybeans. Evidence suggests that a new strain or biotype of western corn rootworms has developed. This new biotype lays eggs in soybeans in the affected area, thus rendering annual crop rotation ineffective.

Currently, more than 25% of the corn acreage in the United States is treated with soil insecticides at planting time to protect plants from attack by corn rootworm larvae. In areas where corn is not rotated, application of soil insecticides is primarily prophylactic. Soil insecticides are applied to nearly 90% of continuous corn acres.

Life History. The western corn rootworm completes 1 generation per year. Females lay eggs from July until the killing frost in the fall; peak oviposition occurs in late August. Most eggs are deposited in the top 4 in (10 cm) of soil, but some can be found as deep as 1 ft (30 cm), depending on the amount of soil moisture during oviposition. Females lay their eggs deeper in the soil in dryland corn or during drought conditions because females reach these depths through soil cracks. Low soil temperatures (19 to 14°F, or –7.5 to –10°C) for 2–3 weeks during the winter reduces viability of eggs and potential for damage during the next growing season. Eggs hatch during 2–3 weeks beginning in late May; peak hatch occurs in early June in much of the Corn Belt.

The three larval instars have head capsule widths of 0.2–0.23 mm, 0.3–0.35 mm, and 0.45–0.5 mm, respectively. When larval development is complete, feeding ceases and a prepupa creates an earthen cell before it pupates; the pupal stage lasts 6–13 days. Rate of development of immatures is temperature dependent. Development occurs fastest between 70–75°F (21–24°C). Within the soil temperature range of 64–81°F (18–27°C), the time from egg hatch to adult emergence is 46 and 23 days, respectively. Depending upon seasonal temperatures, adults begin emerging in early July; peak emergence occurs near August 1. Each female lays as many as 1,000 eggs during her lifetime, but lack of food and freezing temperatures often shorten the adults' longevity.

Gerald R. Sutter

Corn root injury by corn rootworm larvae (Marlin E. Rice).

"Goose necked" corn caused by corn rootworm larvae (Marlin E. Rice).

Injury. The most serious injury is caused by larvae feeding on the roots. Newly hatched larvae feed on root hairs and inside roots; as the larvae mature, they move to the growing tips of roots. If an infestation is heavy, the upper three nodes (primarily nodes 4 through 6) of below-ground roots may be totally destroyed. Destruction of the roots causes plant stress by interfering with uptake of water and nutrients and by reducing the stability of the plant. If plants lodge, yields are further depressed by reduced photosynthetic efficiency and by the difficulty of harvesting poorly anchored stalks.

The extent of yield reduction caused by rootworm damage depends upon environmental conditions. If soil moisture or nutrients are in short supply or if weeds are abundant and competing with corn for these requirements, the effects of larval feeding are increased. The amount of lodging increases if plants are shallow rooted, soils are moist and soft, and strong winds occur. However, plants that have suffered substantial root damage may yield quite well if soil moisture and nutrients are available and injured roots are compensated for by secondary branching of roots.

Injury caused by southern corn rootworm larvae is similar to injury caused by other rootworm species except it occurs earlier in the season, often before the development of brace roots. In some instances, larvae develop early on weed hosts then move to young corn seedlings, damaging the root crowns and causing seedling death. Damage occurs in Texas when corn plants are at the 6- to 9-leaf stages. Larvae chew along roots, excavating grooves and tunnels, and may tunnel into the lower part of the stalk. Severely damaged plants die, resulting in reduced plant stands, or damaged plants are stunted and slower to develop. Yield reduction is correlated with the number of plants exhibiting southern corn rootworm damage and with reduced plant stand.

Corn rootworm larva (Marlin E. Rice).

Newly emerged adult rootworms feed on corn leaves, but they prefer to feed on pollen and silks. Severe silk "clipping" interferes with pollination. If silk clipping is so extensive that less than 1/2 in (1.25 cm) of silks protrude from the husk, pollination may be incomplete. If silking is delayed because of late planting or environmental conditions that slow plant growth, the amount of silk damage and number of barren ears may increase.

Management of Corn Rootworms. *Crop Rotation.* Because Mexican, northern, and western corn rootworm females lay eggs almost exclusively in corn throughout most of their ranges, crop rotation is the most successful method for preventing larval damage. Newly hatched rootworm larvae live only a few days without corn roots to feed on; when a crop other than corn is planted, the larvae starve.

Although crop rotation controls corn rootworms most of the time, exceptions have occurred. For example, when drought conditions prevail and corn plants die during late May, Mexican corn rootworm females migrate into sorghum fields to deposit eggs. Corn planted in sorghum fields the year after a drought can be damaged by rootworm larvae. Other failures of crop rotation to control rootworms have had even more serious consequences.

One explanation for failure of crop rotation to eliminate rootworm problems is that females lay eggs in crops other than corn. This behavior occurs with some frequency when a crop such as soybeans is infested with volunteer corn. However, when a soybean field is free from volunteer corn and other weeds, corn rootworm females usually do not lay enough eggs to result in economic larval damage to corn the next year.

During the early 1990s, the incidence of rootworm damage in first-year corn planted after soybeans in east-central Illinois and northwestern Indiana increased in frequency. Investigation revealed that western corn rootworms caused the damage. Convincing evidence suggests that western corn rootworm females laid eggs in soybeans during the previous season. This situation probably represents an adaptation of western corn rootworms to annual crop rotation. The problem also has been verified in southern Michigan and western Ohio.

Growers in an affected area have relied almost exclusively on crop rotation for rootworm control. The sudden and rather frequent occurrence of rootworm larval injury in corn after soybeans has created concern about the reliability of this cropping practice for rootworm management. Consequently, the use of soil insecticides to control rootworms in corn planted after soybeans in an affected area has increased. However, annual rotation of corn with nonhost crops such as soybeans continues to be a reliable management strategy for corn rootworms in most areas of the United States, and the use of a soil insecticide in this rotation scheme is discouraged.

Another explanation for injury to roots in first-year corn is that northern corn rootworm eggs undergo extended or prolonged diapause, that is, eggs pass through two or more winters before hatching. When extended diapause was first documented (1965), the percentage of eggs with this adaptation was too small to be economically important. However, reports of corn rootworm injury in corn planted after soybeans became more prevalent in the early 1980s, particularly in Iowa, Minnesota, and South Dakota. Diapause in northern corn rootworms is quite variable, ranging from 1 to 4 years. Extended diapause seems to be genetically controlled, suggesting that some populations of northern corn rootworms have adapted to annual crop rotation. Environmental conditions also can affect the expression of extended diapause. Although the extended diapause trait in northern corn rootworms has not been responsible for serious economic damage throughout the Corn Belt, corn in some areas in Iowa and Minnesota is damaged annually by northern corn rootworms that express the trait.

Leaf feeding damage caused by western corn rootworm adults (Marlin E. Rice).

Insecticides for Root Protection. Soil insecticides for rootworm control were first used in the 1950s. By the end of the 1950s, western corn rootworms in Nebraska had developed resistance to the cyclodiene insecticides that were being used. Western corn rootworms became resistant to cyclodienes throughout the Corn Belt by the mid-1960s when granular formulations of organophosphate and carbamate insecticides were introduced. Their use for controlling corn rootworms became one of the largest markets for insecticides for the next several decades.

Most growers who plant corn after corn apply an insecticide to the soil at planting time to protect the roots from larval feeding. Because 4–6 weeks may elapse between the time of insecticide application and egg hatch, and because larval feeding occurs for an additional 3–4 weeks, a product that provides 6–10 weeks of residual activity is ideal. Most corn growers apply a granular formulation in a 7-in (18-cm) band over the row to provide a zone of protection surrounding the primary nodal roots at the base of the plant. Some insecticides can be applied directly into the seed furrow.

Certain insecticides (e.g., carbofuran) have failed to protect corn from larval feeding in some fields because microbes degrade the insecticide at an accelerated rate after they have had a previous exposure to the insecticide. The increased rate of degradation is called enhanced microbial degradation and can occur after a single exposure to the chemical. Avoidance of using the same insecticide every year reduces the potential for enhanced microbial degradation.

Corn rootworm control has been estimated by the level of root protection provided by a soil insecticide. Root protection has been judged on a standard root rating scale of 1 through 6. A rating of 1 indicates few or no larval feeding scars; a rating of 6 indicates that three or more nodes of roots have been pruned to within 1 1/2 in (3.5 cm) of the base of the plant. The estimated root rating at which yield caused by larval feeding is equal to or greater than the cost of applying an insecticide can range from 2.5 to 4.0 (feeding damage equivalent to one node of roots pruned to within 1 1/2 in [3.5 cm] of the base of the plant). Most modern corn varieties have large root systems and have been selected for resistance to drought stress. This trait likely accounts for modern hybrids' ability to tolerate more stress caused by larval feeding; however, hybrid tolerance depends upon environmental conditions.

Because most insecticides are applied in a band over the row, only a small portion of a plant's root system is protected from rootworm feeding injury. Furthermore, roots injured 3–4 in (7.5–10 cm) from the base of the plant may not affect a plant's ability to translocate water and nutrients. Extensive larval feeding also occurs beyond the insecticide-treated band, so larvae survive to adulthood; insecticides reduce survival by only 50%. Consequently, soil insecticides are not good population management tools. However, insecticides may provide measurable yield benefits when rootworm densities are large (at least one or more nodes of roots severely damaged by larval feeding). In essence, soil insecticides protect the brace roots and much of the main root system, thereby preventing stalk lodging and serious yield losses.

In areas with a history of stand loss caused by southern corn rootworms, a registered insecticide applied to the soil at planting time reduces the amount of injury. Economic thresholds for making management decisions within a cropping season have not been developed for this pest. The level of damage caused by southern corn rootworms varies from year to year. Survival of overwintering adults is affected by winter weather, and larval survival is affected by soil moisture and soil type. Little damage occurs during years with very dry or very wet (saturated soil) spring conditions.

Insecticides for Control of Adults. Seed corn and sweet corn growers frequently apply insecticides to prevent corn rootworm adults from clipping silks and interfering with pollination. However, controlling adults also can prevent infestation by larvae the following year if the females are killed before they lay a sufficient number of eggs. This approach for rootworm management, commonly called adult suppression, has both merits and disadvantages.

The primary advantage of adult suppression is that only fields with densities of rootworms above established economic thresholds (about 0.5 beetle per plant) usually are treated with insecticides. However, adult suppression programs require careful and frequent monitoring of plant and insect development to ensure that insecticide applications are timely and that beetles do not migrate from nearby nontreated fields. Several insecticides have been used successfully to suppress corn rootworm populations. However, broadcast applications of broad-spectrum insecticides can be detrimental to nontarget organisms, and may result in outbreaks of secondary pests.

Scouting for rootworm adults involves counting the numbers of beetles on plants (usually 54 per field) several times during August, counting the numbers of beetles captured on yellow sticky cards, or counting the numbers of beetles captured in mechanical traps baited with attractants. The major drawbacks of counting beetles on plants are differences in counts among field scouts and the effects of the time of day, temperature, wind speed, and crop growth stage on beetle activity. The primary problems with the use of yellow sticky cards are difficulty in identifying rootworms among the many trapped insects and locating the traps within a field.

Another approach for adult suppression employs a semiochemical-based bait. Semiochemicals are naturally occurring compounds that modify insect behavior. For suppression of rootworm adults, a small amount of an insecticide is combined with a natural feeding stimulant (cucurbitacin) specific for corn rootworms. The use of this bait reduces the amount of insecticide applied per acre by 95–98%. Adults locate and feed upon droplets of bait and are killed by the insecticide. Proper timing of bait applications have suppressed adult densities to below-threshold levels for the duration of the egg-laying period.

The practice of applying insecticides for conrol of adult corn rootworms has resulted in populations of western corn rootworms in Nebraska that are resistant to methyl parathion and carbaryl. Controlling adult corn rootworms to prevent them from laying eggs needs to be integrated with other management tactics in areas where insecticide resistance has not been detected to avoid placing selection pressure on rootworm populations.

Host-Plant Tolerance. Although no commercial corn varieties suppress larval populations, varieties of corn differ in their ability to tolerate larval feeding. Tolerant varieties have larger root systems and/or a greater ability to compensate for damaged roots. Tolerant varieties do not lodge as readily as more susceptible varieties. Information about hybrids that might be tolerant to rootworm larval injury can be obtained from seed corn companies.

Cultural Practices. Harvesting and planting dates affect subsequent larval populations. If corn is harvested before the end of August (e.g., for silage), significant oviposition during early September can be avoided. Late-planted corn has smaller root systems when eggs hatch, so many larvae starve. Except during years when egg hatch is delayed, a soil insecticide usually is not needed to protect roots from rootworm larval injury if corn is planted after mid-June. Late-planted corn flowers later than corn in surrounding fields. Because adults prefer to feed on silks, they may accumulate in late-planted fields, where the females deposit large numbers of eggs.

Selected References. 46, 74, 143–145, 150–152, 157, 170, 240

Bastiaan M. Drees, Eli Levine, J. W. Stewart, Gerald R. Sutter, and Jon J. Tollefson

Corn Silk Beetle

Corn silk beetle (Marlin E. Rice).

Scientific Classification. *Metrioidea brunneus*, formerly *Calomicrus brunneus*, *Monolepta bifasciata*, *Luperodes davisi*, *Luperodes brunneus*, *Luperus brunneus*, and *Luperodes varicornis*, Coleoptera (Chrysomelidae).

Origin and Distribution. The origin of the corn silk beetle is unclear. It has been found from the Carolinas, along the Gulf Coast to Texas, north to Kansas, and west to Arizona. It also occurs in the Philippines.

Description. The adult is shiny, pale yellow to brown and 1/8–1/6 in (3–4.5 mm) long with an oval, slightly oblong body. The antennae are about two-thirds of the body length. The larva has not been described.

Pest Status. Although not known to be a pest in recent history, complete crop loss was reported frequently as a result of infestations of adults in the Gulf states from 1900 to 1940. During that period, the corn silk beetle also was a pest of cotton, okra, beans, cucurbits, and sweet gum.

Injury. Adults feed on fresh corn silks, sometimes eating the silks all the way to the tip kernels. This injury prevents pollination.

Life History. Little is known about the life history of this beetle. Adults feed on sweet gum early in the spring and migrate to corn and other crops later in the season. In the Gulf states, densities of more than 100 corn silk beetles per ear have caused injury to corn early in July.

Management. Yield loss occurs if silks are continuously clipped back to within 1/2 in (12 mm) of the corn husk before pollination. No insecticides are registered for control of this pest.

Selected References. 25, 80, 112, 124

Michael J. Haas and Douglas A. Landis

Cutworms

Several species of cutworms attack corn in North America. Because each species poses a different threat, they are discussed individually. "Management of Cutworms" applies to all species.

Black Cutworm

Scientific Classification. *Agrotis ipsilon*, formerly *Agrotis ypsilon*, Lepidoptera (Noctuidae).

Origin and Distribution. The black cutworm occurs throughout southern Canada, the continental United States, Hawaii, and Central and South America.

Description. Larvae pass through six or seven instars, depending upon diet and temperature. They can be distinguished from similar species by the convex granules (roughened cuticular texture) on their abdominal segments. Also, the four knotlike tubercles on each abdominal segment along the top of the back are unequal in size. Each pair of tubercles closest to the center of the body is 2–3 times smaller than the pair directly, and more widely spaced, behind them. Larvae are black to pale gray, and their length increases as they grow: third instar, 1/4 in (7 mm); fourth instar, 1/2 in (10–12 mm); fifth instar, 1 in (20–30 mm); sixth or seventh instars, 1 1/2 in (30–45 mm). Pupae are dark brown,

Black cutworm (Marlin E. Rice).

Black cutworm (top) and dingy cutworm (bottom) (Marlin E. Rice).

3/4–1 in (18–25 mm) long and 1/4 in (5–6 mm) wide. Larvae pupate in the soil 3/4–4 in (2–10 cm) deep. The adult has a wingspan of 1 1/8–1 5/8 in (30–42 mm). The basal two-thirds of the forewing is dark, and the outer third is much paler. Each forewing has a dagger-shaped marking at the outer edge of a kidney-shaped spot toward the middle of the wing.

Pest Status. The black cutworm feeds on many hosts, including several weed species. It is the most damaging cutworm species in the Corn Belt, and its damage is particularly severe in late-planted corn after a previous year's crop of soybeans and in corn planted in flood plains. Black cutworms also cause economic damage to vegetables, cotton, tobacco, and turf grasses.

Two other arthropods often mistaken for black cutworms are crane fly larvae and millipedes. Crane fly larvae are not pests of corn, but they are found occassionally in the spring during or shortly after corn planting. The absence of legs distinguishes crane fly larvae from cutworms. Some species of millipedes can injure planted seeds or growing seedlings, especially in no-till corn. However, millipedes rarely cause economic damage. The large number of legs (30 or more pairs, usually two pairs per body segment) distinguishes millipedes from cutworms.

Injury. Economic damage is more common when weed hosts are destroyed at or immediately before corn planting. When weed hosts are killed with cultivation or herbicides, the larvae begin feeding on corn. First to third instars eat small shot-holes in leaves of seedling corn. Fourth to seventh instars feed at night, cutting seedlings at or below the soil surface. Larvae usually consume the cut seedling in a burrow constructed in the soil. As plants become more robust, older instars occasionally tunnel into the growing stalk, disrupting the xylem and phloem. Although climbing behavior in black cutworms is not well developed, larvae will cut off leaves when stalks become too large to cut or too difficult to tunnel. This injury is not economic.

Corn seedlings (coleoptile to 6-leaf stage) cut below ground or larger plants (6–8-leaf stage) that have been tunneled usually do not recover. Seedlings cut above the soil sur-

Crane fly larva (Marlin E. Rice).

Millipede (Marlin E. Rice).

Injury to corn below ground caused by black cutworm drilling into side of stalk (Marlin E. Rice).

Above-ground cutting by black cutworm (Marlin E. Rice).

face often recover if moisture is adequate, but the plants may not yield as well.

Life History. The black cutworm does not overwinter in the Midwest, except occasionally in the Mississippi delta region of southeastern Missouri. Adults migrate from the Gulf states, using low-level wind jets (1,312–5,249 ft [400–1600 m] above ground level), and arrive in the Corn Belt within 2–4 nights. This annual immigration begins in February but is heaviest during April and May. Females deposit eggs singly or in groups (as many as 30) on pasture and fencerow grasses and on low, densely growing weeds and debris in fields that have not been tilled. Damp, low-lying areas in fields that have not been tilled are particularly attractive for oviposition. Females firmly attach eggs to densely growing plants such as curled dock, yellow rocket, velvetleaf, and rough pigweed, or to fine-textured plant debris such as dead foxtail or soybean stems. Females rarely deposit eggs on living soybean or corn plants.

Minimum developmental threshold is 50°F (10°C) and degree-day accumulation can be used to predict larval feeding activity. Degree-day accumulation begins when the first significant adult flight occurs (see "Management Tactics for Cutworms"). Plant cutting begins with fourth instars and is estimated at 300 FDD (167 CDD). Pupation occurs at approximately 641 FDD (356 CDD).

Except for late-planted sweet corn, corn in the Corn Belt is injured only by the spring generation. Females of the spring generation (June or July) may stay in the area to deposit eggs in pastures, fencerows, vegetable fields, and gardens, or move farther north on persistent low-level wind jets. Females in the summer generation (July–August) deposit eggs on pasture and fencerow grasses and on turfgrasses. The autumn generation (September–October) occurs in similar habitats as the previous generations. Larvae reside 1–4 in (2–10 cm) deep in soil where moisture concentration is higher. During extremely wet conditions, however, larvae climb on vegetation.

From late August through October, adults respond to changing environmental conditions and fly south, assisted by surface winds, to the coastal regions of the Gulf of Mexico. Black cutworm pupae survive the winter in soil as far north as the 32°F (0°C) isotherm. This line fluctuates annually in the mid-continental United States but rarely is farther north than central Tennessee, southeastern Missouri, and central Oklahoma.

The most important parasitoids that attack older black cutworm larvae are the fly *Bonnetia comta* and the tiny wasp *Meteorus vulgaris*. Predation by ground beetle larvae and adults (*Scarites substriatus*, *Pterostrichus chalcites*, *P. lucublandus*, and *Harpalus pennsylvanicus*) also regulates populations of black cutworms.

William B. Showers

Bristly Cutworm

Scientific Classification. *Lacinipolia renigera*, Lepidoptera (Noctuidae).

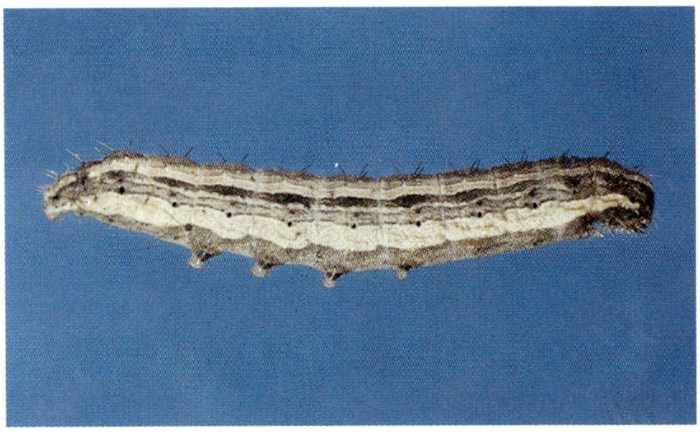

Bristly cutworm (Roy W. Rings).

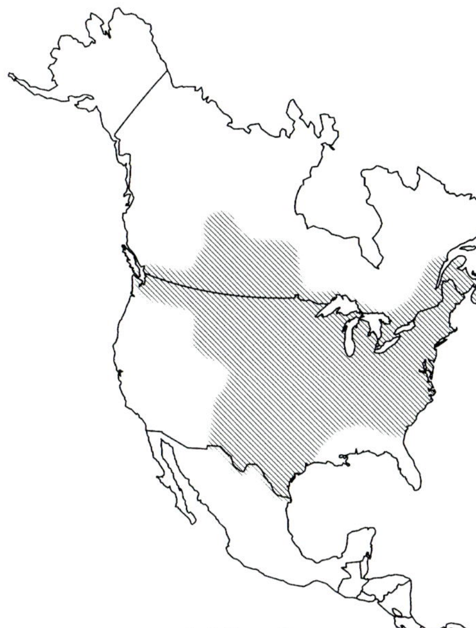

bristly cutworm

Origin and Distribution. The bristly cutworm is found throughout the United States east of the Mississippi River. Its western range includes Colorado, Kansas, Louisiana, Nebraska, New Mexico, North Dakota, and Texas. This species also is found in Canada and Europe.

Description. The common name of this pest refers to the stiff hairs that protrude from all parts of the larva's body, which is dull gray-brown with prominent stripes along the sides. The back of the larva has a pale design of inconspicuous dark, diamond-shaped markings. A full- grown larva is about 1 in (25 mm) long.

Pest Status. The bristly cutworm is a minor pest of corn, although it occasionally causes heavy damage.

Life History. The bristly cutworm completes 2 generations per year, and the second generation overwinters as partly grown larvae. The first generation occurs from late May to late July, with peak adult abundance in June. The second generation occurs from late August to late October, with peak adult abundance in September. The larvae feed near the soil surface and can survive on many plants, including fruit and vegetable crops, corn, clover, tobacco, and some weeds. However, bristly cutworms feed mostly on noncultivated plants or on hay, grasses, and legumes. Because feeding injury is not obvious on these crops, management of this cutworm is seldom necessary.

Armon J. Keaster

Bronzed Cutworm

Scientific Classification. *Nephelodes minians*, formerly *N. emmedonia*, *N. expansa*, *N. sobria*, and *N. subdolens*, Lepidoptera (Noctuidae).

Bronzed cutworm (Marlin E. Rice).

Origin and Distribution. This species usually is found between 35°N (North Carolina, Tennessee, Missouri) and 50°N (Newfoundland to British Columbia) latitude. However, it has been found in every state except the Gulf states, Utah, and Wyoming.

Description. The larva has six instars and is 1 1/2–1 3/4 in (35–42 mm) long when full grown. It is dark brown with a bronze sheen and has four brown and three yellow stripes extending the length of the body.

Pest Status. This species is an important pest of bluegrass and other turfgrasses. Economic problems in corn occur when it is planted into sod or pasture grasses. The bronzed cutworm also feeds on buds and leaves of fruit trees.

Injury. Larvae feed deep in the crowns of grasses and at the soil surface on young corn plants.

Life History. The bronzed cutworm completes 1 generation per year. Females deposit eggs on grasses or debris from late September to early October. Eggs remain in diapause through the winter and hatch in the spring. Larvae are full grown by early May, burrow into the soil, and become dormant until pupation begins in mid-August. Adults emerge in mid-September to early October, mate, and females lay eggs.

William B. Showers

Claybacked Cutworm

Scientific Classification. *Agrotis gladiaria*, Lepidoptera (Noctuidae).

Origin and Distribution. The claybacked cutworm is a native of North America and occurs from Nova Scotia to British Columbia, throughout the Midwest, and from Quebec to Florida. It is most abundant in the north-central and eastern United States.

Description. A broad, yellow-brown stripe on the back

Claybacked cutworm (Roy W. Rings).

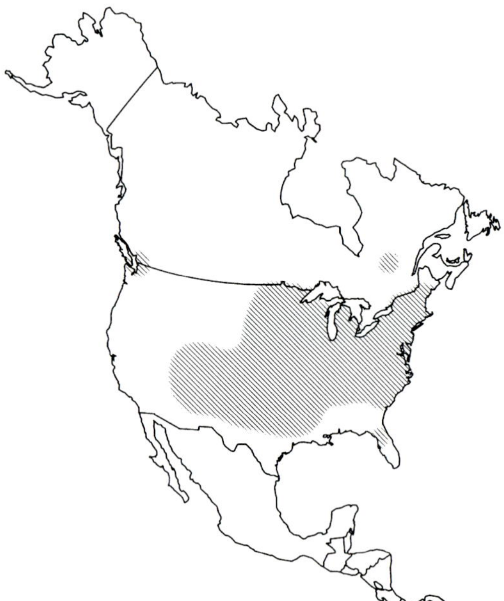

claybacked cutworm

of the larva explains the common name of this cutworm. The rest of the body is pale gray and translucent, and the head is gray-brown with bars on the front of the face. A full-grown larva is 1 1/3 in (33–37 mm) long.

Pest Status. The claybacked cutworm can be very destructive to seedling corn. Serious local outbreaks occur sporadically in the eastern half of the United States. The claybacked cutworm does not occur in damaging numbers every year.

Injury. Small claybacked cutworm larvae pull the corn leaf down into a burrow and eat the leaf from tip to base. Larger larvae cut entire leaves or plants at or just above the soil surface and drag them into the burrow where feeding continues underground. The larvae are active at night and remain in their burrows during the day. A newly germinated field infested with claybacked cutworms appears as if it were broadcast seeded because of the cut plants jutting from the burrows between the rows. Damage caused by claybacked cutworms often is confused with damage caused by black cutworms because their appearance and feeding habits are similar.

Life History. Claybacked cutworms complete 1 generation per year. Partly grown larvae overwinter and resume feeding in early spring. Larvae complete development by late May or early June in the Corn Belt. Mature larvae enter the soil and construct pupation cells but do not transform into pupae for 6–8 weeks. Adults emerge during September or early October. Females deposit eggs around the base of favored host plants such as clover, corn, tobacco, vegetables, strawberries, blackberries, raspberries, and weeds. Newly hatched larvae feed as long as hosts are green, and then they become dormant during the winter. The cycle is repeated when the overwintering larvae begin to feed again in the spring.

Armon J. Keaster

Dingy Cutworm

Scientific Classification. *Feltia ducens*, Lepidoptera (Noctuidae).

Origin and Distribution. The common name "dingy cutworm" was shared by several species (*F. ducens*, *F. subgothica*, *F. jaculifera*, and *F. herilis*) until 1978. Because of this confusion, the distribution of each of these species cannot be established precisely. *Feltia ducens* occurs in southern Canada southward through Virginia, Tennessee, Missouri, Kansas, Colorado, and Utah.

Description. Eggs are white, oblong-oval, and flat in profile. The upper half of the egg turns brown 24 hours after being deposited. Larvae pass through 6 instars. Third to sixth instars are 1/4–1 1/4 in (9–32 mm) long. Larvae are pale gray to brown, tinged with red. A faint, dark, V-shaped marking

Dingy cutworm (Marlin E. Rice).

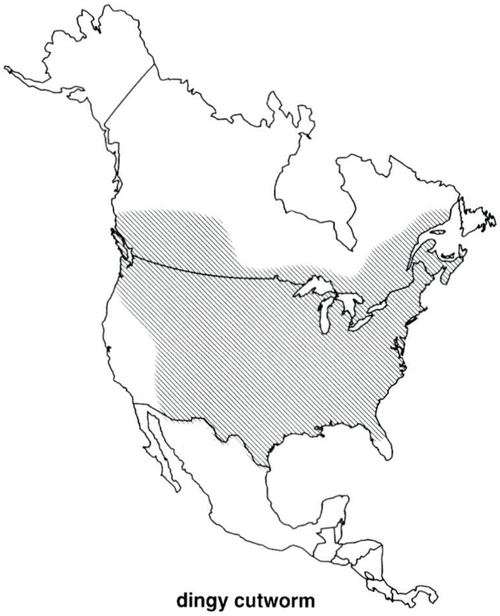

Glassy cutworm (Marlin E. Rice).

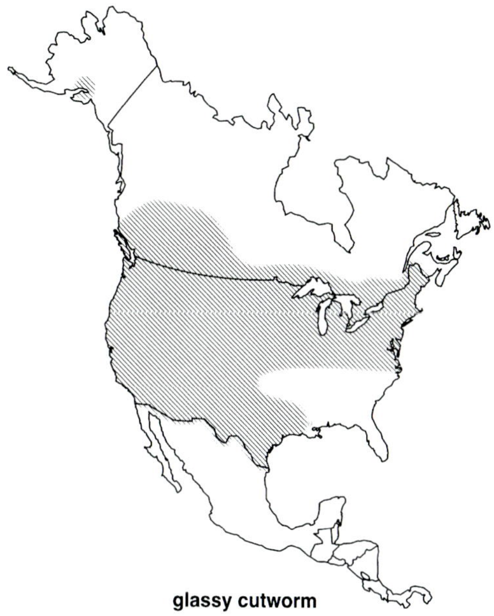

appears on the back of each abdominal segment. The head is pale brown-gray. Tubercles along the top of the abdominal segments are equal in size, in contrast to the unequal-sized tubercles on the back of the black cutworm. Tubercles become larger, however, toward the posterior segments. The pupa is 3/4 in (18 mm) long.

Pest Status. Dingy cutworms feed on vegetables, clover, alfalfa, tobacco, wheat, corn, grasses, and broadleaf weeds. They are infrequent pests of corn.

Injury. The larvae nip the ends of young corn leaves, and only rarely cut or drill into young corn plants.

Life History. The dingy cutworm completes 1 generation per year. Egg hatch occurs in late August or September. Larvae feed and develop to the third or fourth instar by October when they enter diapause. Larvae break diapause during March or April and feed. Fourth instars are present in cornfields at planting time, particularly in weedy fields or in corn planted into pasture or alfalfa. Larvae become dormant in mid-May, break dormancy during July, and begin pupating from late July to early August. The larva pupates 1/2 in (1 cm) below the soil surface. Adults emerge during mid-August to September and mate. Females lay eggs on leaf sheaths of plants growing low to the ground. The most important parasitoid is the small wasp *Microplitis feltiae* that parasitizes about 20% of a population of dingy cutworms.

William B. Showers

Glassy Cutworm

Scientific Classification. *Apamea devastator*, formerly *Crymodes devastator*, Lepidoptera (Noctuidae).

Origin and Distribution. The glassy cutworm occurs throughout North and South America and is widespread in the United States, except in the southern states.

Description. This cutworm gets its name from its green-white body that appears translucent or glassy. The head and pronotum are red-brown. Full-grown larvae are 1 1/2 in (35–40 mm) long.

Pest Status. This species prefers sod as a host, so it is more often a pest in crops planted after sod or pasture, especially in low ground. The glassy cutworm also attacks almost all vegetable crops but prefers beans, cabbage, and corn. It occurs infrequently at outbreak densities.

Injury. This sod-infesting cutworm is found in subterranean burrows, feeding on roots and basal stems of corn and other grasses.

Life History. The glassy cutworm overwinters as small larvae, and the larvae complete their development in the spring. One generation occurs annually. Adults fly from late

June to early September. The larva's behavior and habitat are more subterranean than that of most other cutworm species.

Armon J. Keaster

Pale Western Cutworm

Scientific Classification. *Agrotis orthogonia*, formerly *Porosagrotis orthogonia*, Lepidoptera (Noctuidae).

Origin and Distribution. The pale western cutworm is native to the Great Plains of North America. It occurs in western Manitoba south to the Texas panhandle and westward through the Rocky Mountains.

Description. Eggs are white when first deposited, but they soon become yellow-gray. They are spherical and about 1/16 in (1.6 mm) in diameter. Females deposit eggs in clusters of 30 or 40 eggs, usually 1/4–1/2 in (7–10 mm) below the surface of soft, loose soil. The larva has seven instars and changes from yellow-brown to slate gray with three pairs of green-gray stripes along the back and sides. The cuticle is semitransparent, so internal organs are visible through the cuticle. The head is amber to black. Black markings on the front of the head resemble a narrow-waisted **H** on young larvae and an inverted **V** on mature larvae. Mature larvae are 1 1/4–1 1/2 in (3–4 cm) long. The dark brown pupa is about 5/8 in (1 cm) long.

Pale western cutworms (Entomological Society of Saskatchewan).

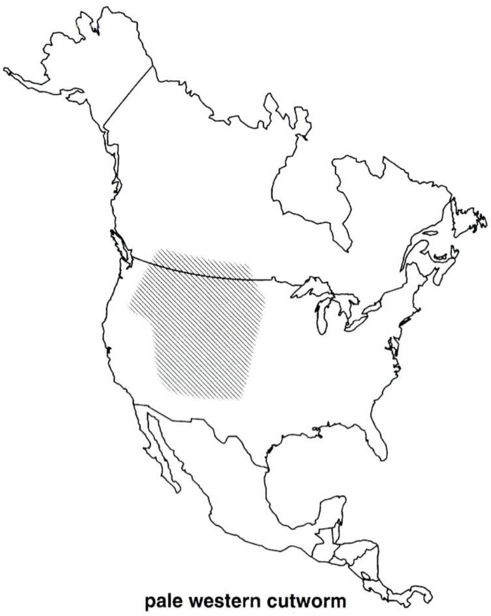

pale western cutworm

Pest Status. This species feeds on wheat, oats, corn, barley, grass, beans, alfalfa, sunflower, and weeds. Feeding preference is dictated more by soil texture and moisture (dry is preferred) than by host plant species. The pale western cutworm causes periodic damage throughout its range on dryland vegetation. Outbreaks occur most frequently in the dry valleys within the Rocky Mountains.

Injury. The pale western cutworm burrows through the soil feeding on the stems 1/4–1 in (6–25 mm) below the soil surface. Seedling plants wilt and die. While foraging, larvae usually follow the row where the soil is looser, cutting consecutive plants. Unless soil moisture is considerable, this cutworm occurs on the soil surface for only a few days after hatch. Thereafter, it burrows into the soil to feed and develop. The larvae seek soil moisture 1–6 in (2.5–15 cm) below the surface. The greatest amount of injury usually occurs from April to June in dryland crops in low-moisture environments. Irrigated crops are not susceptible to attack by this cutworm.

Life History. The pale western cutworm completes 1 generation per year. In the southern part of its range, the adult emerges from the soil in late August; flight and mating continue until frost in late October. In the northern part of its range, adult emergence also begins in late August, but flight and mating cease by mid-September. Females usually lay eggs in loose soil during late afternoon before sunset. The embryo matures into a fully formed larva within several days, then the unhatched larva enters diapause. Eggs do not hatch until January (southern range) or April (northern range). Larvae feed throughout the spring, after which time mature larvae burrow deeper into the soil, produce an earthen cell, and become dormant. Pupation occurs in late July or early August.

Dry seasons, particularly during oviposition, favor an increase in densities of this cutworm. Cool, wet weather when females are laying eggs and early instars are feeding (January to May) dramatically reduces population density.

Very little parasitism of this subterranean species occurs. The wasp *Meteorus vulgaris* occasionally parasitizes older larvae.

William B. Showers

Redbacked Cutworm

Scientific Classification. *Euxoa ochrogaster*, formerly *Nocta ochrogaster*, Lepidoptera (Noctuidae).

Origin and Distribution. The redbacked cutworm occurs throughout Canada, from coast to coast, and south to Colorado.

Description. In Canada, redbacked cutworms often occur with other species of the same genus. Most of the species have similar feeding habits, so not all crop damage attributed to the redbacked cutworm is caused by this species. Mature larvae are 1 1/2 in (36–40 mm) long and light brown to gray. Two broad, red bands border a light, median stripe on the back.

Redbacked cutworms (Lloyd Harris).

Pest Status. The redbacked cutworm is the most widely distributed and destructive cutworm in Canada, attacking corn, wheat, oats, barley, vegetables, sunflowers, sugar beets, and ornamental flowers. Because eggs are deposited in the field in the fall, predicting the location of an infestation and the crops that will be attacked is difficult. However, an infestation can be detected early by looking for evidence of larval feeding on weeds or on crops.

Injury. Early instar larvae climb plants and feed on leaves, but feeding by young larvae accounts for less than 6% of total plant injury. Late-instar larvae cause most of the economic damage by severing plant stems at ground level.

Life History. The redbacked cutworm completes 1 generation per year and overwinters as an egg. Eggs complete embryonic development and enter diapause. Larvae emerge the following spring, usually in April, and feed for 6–8 weeks, mostly at night. Most crop damage occurs in June. In Manitoba, adults begin to emerge in early July; peak flight occurs in early to mid-August. In July and August, females deposit eggs in the soil.

William B. Showers

Sandhill Cutworm

Scientific Classification. *Euxoa detersa*, Lepidoptera (Noctuidae).

Origin and Distribution. The sandhill cutworm is distributed from Nova Scotia and Quebec to Washington and south to South Carolina and Colorado. Its occurrence seems to be limited to sandy soils.

Description. The larva is white to pale gray. Pulsations in the blood vessel along the back can be seen through the cuticle of live larvae. Faint, chalky-white stripes are evident on the back and sides, and the head is dull red-brown. A full-grown larva is about 1 1/3 in (30–35 mm) long.

Pest Status. This cutworm generally is not a problem in corn in the Midwest, except in sandy soils throughout its range. The sandhill cutworm also has been recorded as a pest of peas, melons, oats, onions, rape, raspberry, rye, strawberry, sweet clover, and wheat.

Injury. Sandhill cutworm larvae move below the soil sur-

Sandhill cutworm (Marlin E. Rice).

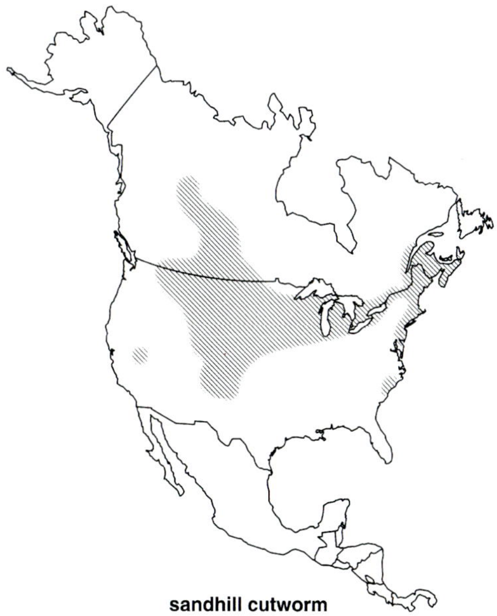

sandhill cutworm

face and feed only on underground parts of the plant, sometimes cutting stems. However, the larvae seldom are seen because they spend most of their time underground.

Life History. Sandhill cutworms occur almost exclusively in sandy soils. They overwinter as partly grown larvae. Larvae resume feeding in May and become full grown by early June. Mature larvae pass the hot summer in underground cells and pupate in early August. Adults emerge and are present during late August and September. Sandhill cutworms complete 1 generation per year.

Armon J. Keaster

Spotted Cutworm

Scientific Classification. *Xestia* spp., formerly *Amathes c-nigrum*, Lepidoptera (Noctuidae).

Origin and Distribution. Spotted cutworms generally are distributed throughout North America. They also are found in Europe and Asia.

Description. Spotted cutworms are dull gray-brown, with a broad darker stripe along each side of the lower part of the body. A series of black, wedgelike markings begin partway along the larva's back and are larger toward the rear, giving the larva a spotted appearance. A full-grown larva is about 1 1/3 in (35 mm) long.

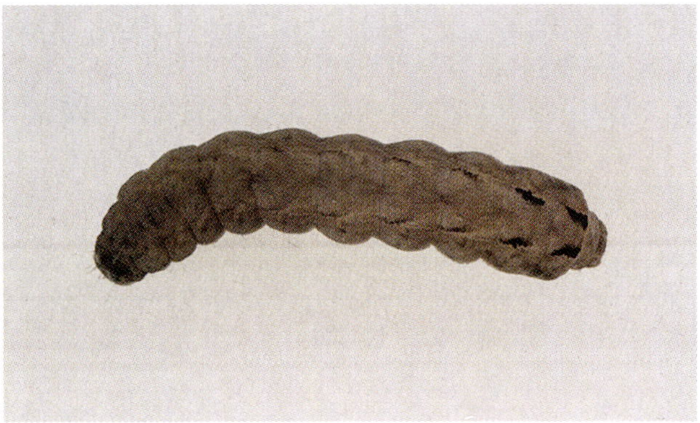

Spotted cutworm (Agriculture and Agri-Food Canada).

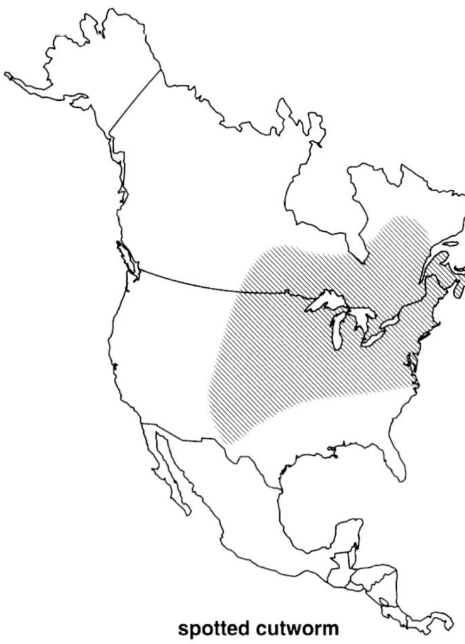

Pest Status. The spotted cutworm is common but is of minor economic importance in the Midwest. It is considered to be an important pest of gardens in the northeastern United States.

Injury. The spotted cutworm is a climbing cutworm and climbs the host plant to feed on leaves, stems, buds, or fruit. These cutworms feed at night. They also occasionally climb fruit and shade trees at night.

Life History. Spotted cutworms complete 2–3 generations per year. They overwinter as large larvae. Adults are most abundant in early to mid-June and in late August or September. Females deposit eggs singly, in rows, or in patches on the leaves of host plants. Although this cutworm prefers vegetable crops, it is a generalist feeder and also feeds on fruit trees, cereal crops, tobacco, and many other crops.

Armon J. Keaster

Variegated Cutworm

Scientific Classification. *Peridroma saucia*, formerly *Lycophotia saucia*, Lepidoptera (Noctuidae).

Origin and Distribution. The variegated cutworm occurs in every Canadian province, the continental United States, and south to Argentina.

Description. Females deposit eggs in a single layer, as many as 100 at a time. Larvae develop through six instars. Although the larva's color varies considerably, a narrow line of pale yellow dots along the middle of the back is almost always present. Third to sixth instars are 1/4–1 1/2 in (4–40 mm) long. Pupae in the soil are 3/4–1 in (15–25 mm) long.

Pest Status. Variegated cutworms attack fruit and forest trees; garden vegetables; ornamental flowers; and field crops, including corn, alfalfa, clover, cotton, sunflower, tobacco, and wheat. These cutworms cause economic damage most often to potatoes and tomatoes.

Injury. Larvae feed on the leaves of corn and may eat the center of the stem down to 2–3 in (5–8 cm) below ground.

Life History. Variegated cutworm adults probably use low-level wind jets, 1,312–5,249 ft (400–600 m) above ground, in the spring to migrate from coastal regions of the Gulf of Mexico into the Corn Belt. They probably use northerly surface winds to return to the Gulf Coast in the fall. Females deposit eggs in pastures; fencerow grasses; low, densely growing weeds; and debris in fields that have not been tilled. The white, circular eggs mature in 2–4 days. Larvae develop

Variegated cutworm (Marlin E. Rice).

through 6 instars, feeding primarily at night. Early in the morning, larvae drop onto the ground where they burrow into the soil and remain until evening. Two to 4 generations occur in North America north of Mexico, depending upon temperature and location. Corn in the Corn Belt is damaged only by the spring generation during May and June.

Several parasitoids attack variegated cutworm larvae. Wasps (*Meteorus vulgaris*, *M. autographae*, and *M. indagator*) and flies (*Archytas analis*, *A. apicifera*, and *Phorocera saundersii*) are the most important parasitoids. As many as 50% of pupae may be parasitized by the wasp *Ichneumon capitus*. The ground beetle *Scarites subterraneus* preys upon variegated cutworm larvae. Birds also feed on the larvae.

William B. Showers

Management Tactics for Cutworms

Cutworms continue to challenge the best efforts at pest management because of their irregular distribution in location and time. Cultural tactics, trapping of adults and larvae, scouting, and insecticides can be used for management of cutworms. Cultural tactics that reduce the potential for the occurrence of some species of cutworms include elimination of winter and early spring vegetation before planting and promotion of rapid seedling growth. Removal of cool-season weeds with cultivation or herbicides at least 1–2 weeks before planting, especially in no-till fields, starves small cutworm larvae before corn emergence. Because cutworms usually damage seedlings smaller than the 6-leaf stage, any practice that promotes rapid seedling growth reduces the potential for cutworm injury. When possible, growers should avoid planting corn in fields with a history of cutworm problems and should avoid planting corn after long-standing pastures, meadows, alfalfa, or red clover.

Trapping adults and larvae may be an important tool for determining the potential threat posed by some species of cutworms. Several states in the Corn Belt have a coordinated black cutworm trapping program for adults. Traps are baited with the synthetic sex pheromone of black cutworm females. During the spring, many of the migrating females mate during the flight from the Gulf of Mexico coastal areas. Upon arriving in the Corn Belt, males search for virgin females and are attracted to the baited traps. A significant flight is indicated when eight males are captured in a 2-night period. The date of this significant capture is used to predict larval development and the date of first cutting injury by black cutworm larvae in cornfields. Scouting for cutting of seedling corn should begin when 300 FDD (167 CDD) have accumulated after a significant trap capture.

Entomologists at the University of Missouri have developed a larval cutworm bait trap that can be used to assess densities of several species of cutworms. The cutworm trap, containing wheat seedlings, wheat bran, and a green screen cylinder, attracts black, bristly, claybacked, dingy, sandhill, spotted, and variegated cutworms. The economic thresholds based upon captures of larvae in these traps have been developed.

The entomophagous nematode *Steinernema carpocapsae* shows promise for suppressing densities of black cutworm larvae.

Insecticides applied before or at planting or before corn emergence can be used to prevent cutworm damage. However, the management tactic most often recommended involves scouting for cutworms and their injury and application of an insecticide only after the number of plants injured has exceeded established levels. During years when outbreaks occur, cutworms are difficult to control even with insecticides.

The decision to apply a postemergence insecticide should be based upon the density of cutworms and the associated level of injury. Scouting is extremely important to detect cutworm problems that might be developing within each field. Scouting for cutworm feeding injury should begin as soon as corn emerges from the soil. Control with postemergence insecticides is warranted when 6–8% of seedling plants are being fed upon or cut above ground or when 2–4% are cut below ground. The lower percentage of cutting injury is recommended if the plant population is low; the higher percentage of cutting injury is recommended when the plant population is high.

Some *Bt* corn hybrids offer partial control of black cutworms by suppressing their populations. Refer to the "European Corn Borer—Management" section (p.81) for more details about transgenic corn and resistance management strategies.

Selected References. 4, 14, 37, 48, 53, 55, 56, 86, 96, 111, 176, 182, 202–212, 221, 222, 228, 234, 253

Armon J. Keaster

European Corn Borer

European corn borer larva (Marlin E. Rice).

Scientific Classification. *Ostrinia nubilalis*, formerly *Pyrausta nubilalis*, Lepidoptera (Crambidae).

Origin and Distribution. The European corn borer was introduced into North America from Europe in the early 1900s, possibly in infested broom corn from Italy. Since its introduction, this moth has adapted to many hosts and environmental conditions. It occurs from the Atlantic coast to the Rocky Mountains and from the Gulf of Mexico to southern Canada. Its host range consists of more than 200 species of plants, including many common weeds, and it is a pest of many agronomic and horticultural crops, including snap beans, potatoes, peppers, apples, small grains, and cotton. Depending on latitude and elevation, the corn borer completes 1–5 generations each year, so the seasonal distribution of each life stage varies throughout its range.

Description. Egg masses are about 1/4 in (6 mm) in diameter, are flat, and contain from a few to as many as 60 eggs (average 23). A single egg is about the size of a pinhead, and the eggs of a mass overlap like fish scales. A freshly deposited egg mass is creamy white to light green but becomes transluscent as it ages. Within 24 hours of deposition, individual eggs appear to have a black center. This stage is called the blackhead stage because the black centers are the dark brown to black head capsules of the larvae developing within.

Newly hatched larvae are 1/32–1/16 in (1–2 mm) long with a dark brown head and a somewhat translucent white body. Mature larvae are 3/4–1 in (19–25 mm) long with a medium to dark brown head and a creamy white to gray body. Raised, sometimes slightly darkened tubercles are evident on the body. All European corn borer larvae have three pairs of true legs just behind the head and five pairs of prolegs on their abdomen.

Pupae are 1/2–3/4 in (12–19 mm) long, medium to dark brown, with a rounded head region and an abdomen tipped with a tiny hook.

The adult female is about 1 in (25 mm) long. The female's wing pattern is an alternation of yellow-brown and medium brown wavy lines across the wings. When the eggs are fully developed, the female's abdomen is swollen and extends beyond the length of the wings. Males have a similar but darker wing pattern. The abdomen of the male always extends beyond the wings, is narrow, and has a tufted, hairy tip.

Pest Status. The European corn borer is a frequent economic pest of field corn. However, because of corn's ability to tolerate some feeding injury, large fluctuations in corn borer densities from year to year, and lack of scouting efforts, the impact of European corn borers on corn yield often is underestimated. In higher-value crops such as seed corn, sweet corn, and popcorn, for which protection of the ear is important, the European corn borer frequently is economically damaging. The corn borer also is potentially more serious in areas where the first generation develops on one host and the next generation attacks nearby corn. In general, economically damaging populations of corn borers develop more frequently in the western portion of the Corn Belt.

Injury. Corn borer larvae feed on all parts of the corn plant except the roots. When larvae hatch during the pre-tassel stage of corn growth, they feed primarily within the whorl and also tunnel into leaf midribs. Larger larvae (third instar and larger) tunnel into the lower stem. Whorl-feeding injury to the leaves seldom causes significant yield loss. After corn tassels have emerged, small corn borer larvae are found in the leaf collar and sheath areas or in silk channels; larger larvae (third instar and larger) tunnel into the stalks, ears, or ear shanks. In sweet corn, seed corn, and popcorn, larvae feed on

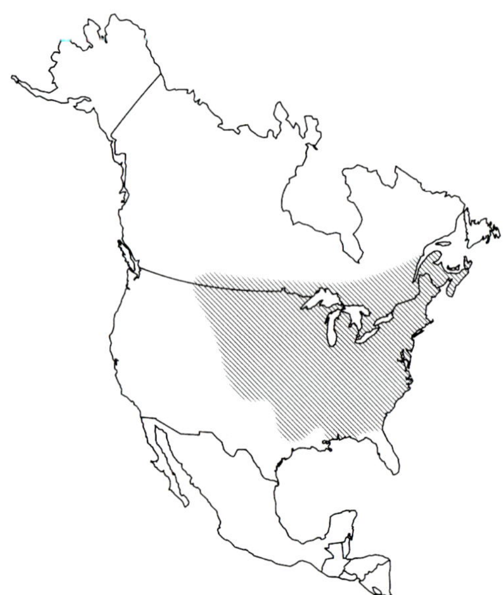

European corn borer

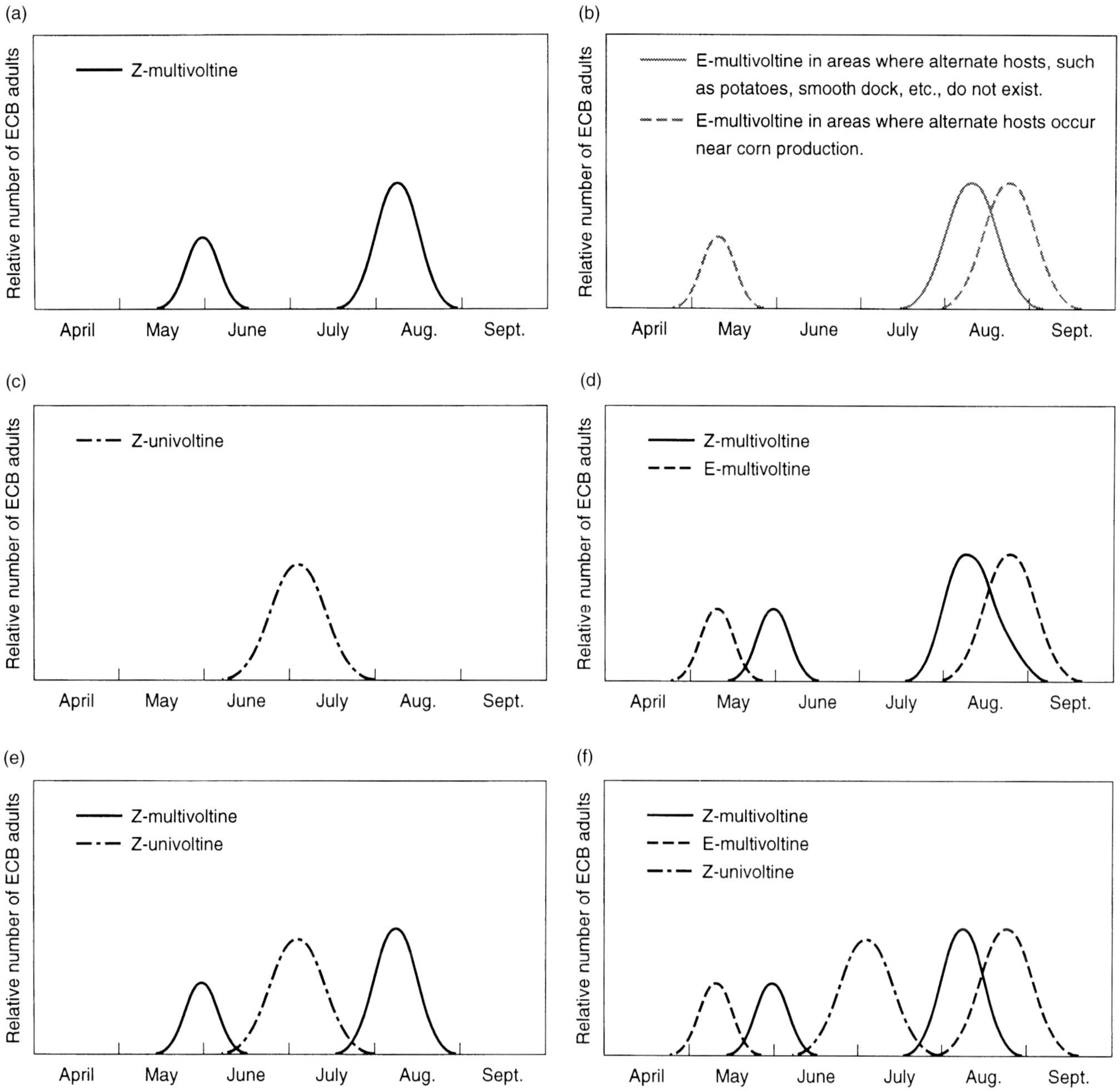

Fig. 10 a–f. Adult European corn borer flight periods: (a) Z-multivoltine population; (b) E-multivoltine population; (c) Z-univoltine population; (d) Z- and E-multivoltine populations co-existing; (e) Z-multivoltine and Z-univoltine populations co-existing; (f) all three populations co-existing.

the developing seeds.

The effects of tunneling on the plant's ability to move nutrients, minerals, and water often reduce grain weight and kernel number, particularly when tunneling occurs before or during the early reproductive stages of the plants. Yield reductions also result from stalk breakage and ear drop. Pathogens that enter through corn borer feeding wounds may cause stalk rot, resulting in additional yield loss.

Life History. The life history of the European corn borer is influenced by local conditions and the genetics of the insect. In some northern latitudes and at higher elevations, a single-generation (univoltine) type occurs and can coexist with a two-generation (bivoltine) type. In other areas, 2 or 3 multiple-generation (multivoltine) types can coexist, with each type expressing a slightly different life history. In addition, corn borer males in different areas respond to two different ratios of the female's sex pheromone and have been designated Z and E strains.

Figure 10a shows typical adult flights during May and June (first generation) and during July and August (second

European corn borer eggs (left), European corn borer eggs in blackhead stage (right) (Marlin E. Rice).

generation) in the Corn Belt (Z-multivoltine population). Figure 10b presents adult flights for areas along the upper East Coast (e.g., New Jersey) and for an E-multivoltine population. This population's first flight typically occurs slightly earlier (late April and early May) and the second flight typically occurs slightly later (mid-August to late September) than the corresponding generations in the Corn Belt. A single-generation type of flight period is shown in Fig. 10c (Z-univoltine population). Figures 10d–f represent European corn borer flight periods in areas where two or three types coexist.

All European corn borers overwinter as mature larvae within corn stubble, corn ears, stems of other hosts, or plant debris in or on the soil. The differences in timing of the first flight of each type of European corn borer occur because of both the length of time spent as an overwintering larva before pupating and the local weather conditions that affect development. All types spend from 7 to 10 days in the pupal stage. Adults from overwintered caterpillars emerge and females lay eggs for the first generation in cornfields or on other crops or wild hosts, typically on the undersides of leaves. Hatching occurs in 5–7 days when temperatures are favorable, ranging from 10 to 4 days at 65–90° F (32–18°C). Hatched larvae move immediately to the whorl, ear, or leaf collar depending upon the growth stage of corn. Larvae may remain in these sites, or, after reaching the third or fourth instar, tunnel into the stem or leaf midrib where they develop and pupate.

First-generation adults emerge 7–10 days after pupation and congregate in grassy areas (where the humidity is high) along field edges or within fields, especially in the Corn Belt. The adults also may move into soybean or alfalfa fields, or into other areas of dense vegetation. In some southern areas, adults congregate in cornfields. These areas of congregation are called "action sites" because large numbers of adults accumulate in a relatively small area to mate and rest. Three to 5 days after emergence, females begin to lay eggs for the second generation. Corn at this time typically is in the reproductive stages of development. Females are attracted to cornfields in which the plants are silking and shedding pollen. About 90% of all eggs are deposited on the undersides of the ear leaf and on the three leaves above and below the ear leaf. The second and subsequent generations pass through the same series of stages as the first generation. In the Corn Belt and at similar latitudes, most second-generation, fifth instars enter diapause to overwinter. In the Mid-Atlantic area, second-generation larvae may not diapause in years of substantial heat accumulation; farther south, few second-generation larvae diapause. Diapause is induced by a combination of decreasing day length, temperature, and quality of food.

Newly hatched European corn borer and feeding injury (Marlin E. Rice).

Whorl feeding by Eurpoean corn borer larvae (Marlin E. Rice).

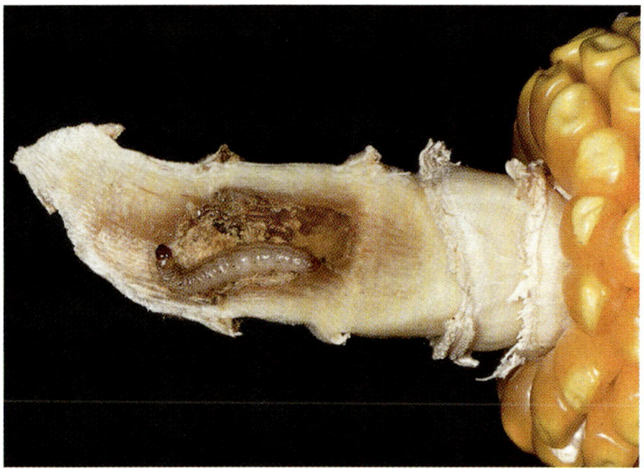

European corn borer larva in ear shank (Marlin E. Rice).

Scarred kernels from European corn borer feeding (Marlin E. Rice).

Subsequent generations of European corn borer may attack late-maturing corn or other cultivated or wild hosts and remain at these sites as overwintering larvae.

The European corn borer has several natural enemies. Insect predators include the minute pirate bug, damsel bugs, several species of lady beetles, lacewings, ground beetles, bigeyed bugs, and other general predators. Vertebrate predators include birds, bats, and other rodents. Insect parasitoids include *Trichogramma* spp., *Lydella thompsoni*, *Eriborus terebrans*, *Macrocentrus grandii*, *Simpliesis viridula*, *Chelonus annulipes*, *Campoletis* sp., *Archytas marmoratus*, and *Lixophaga* sp. Pathogens of the European corn borer include *Bacillus thuringiensis*, *Nosema pyrausta*, *Beauveria bassiana*, *Metarhizium anisopliae*, *Hirsuttella* sp., and *Aspergillus niger*. These organisms may exert some influence on populations of European corn borers.

Management. Because adult European corn borers are very mobile, crop rotation and tactics that isolate the crop do not manage this pest. Destruction of overwintering larvae by stalk shredding or plowing reduces the local abundance of first-generation caterpillars in early spring crops such as field corn, sweet corn, and potatoes. However, stalk destruction has little impact over a wide area and little influence on later generations. Resistant hybrids, early crop maturity (short-season hybrids and early planting), biological control, early harvest, monitoring for adults and egg masses, and chemical and biological (*Bacillus thuringiensis* var. *kurstaki*) insecticides are tools used to manage this pest.

Pheromone traps can be used to monitor adults to determine the best times for scouting. These traps must be baited with an effective, commercially available lure of the correct pheromone blend for the predominant local strain, either the Iowa (Z) or the New York (E) strain. In some states, both strains may occur in separate or overlapping areas. The *Heliothis* (Hartstack) trap is the most effective commercial trap available. Pheromone traps should be situated either in or near action sites along field margins or, for pollinating corn, within the field at ear level. Scouting in reproductive-stage corn should be started about 5 days after trap catches increase sharply.

The earliest planted cornfields in an area are particularly susceptible to attack by the first generation. Fields that are pollinating when adults are flying may be susceptible to either first- or second-generation borers, depending upon latitude. Very late-maturing fields also may be susceptible to attack by later generations of the pest. These fields should be monitored closely during periods of adult activity to identify a developing pest population.

Estimating the percentage of whorl-stage plants with signs of injury and counting the number of larvae in some infested plants are recommended for scouting first-generation European corn borers. Scouting for second-generation borers involves examining leaves in the ear zone for egg masses. Insecticides may be warranted when published thresholds are exceeded. Linker et al. (154), Mason et al. (165),and Tollefson and Calvin (245) describe scouting protocols and thresholds for first- and second-generation European corn borers in different areas of the United States. Chemical control in whorl-stage corn can be very effective, but insecticide applications for control of second-generation larvae often are only moderately effective unless they are timed and applied properly. Aerial applications of granular insecticides usually are more effective for controlling first-generation corn borers than aerial application of liquid insecticides.

Gene transfer techniques have been used to produce corn plants that contain a gene taken from *Bacillus thuringiensis (Bt)*. The *Bt* gene has been inserted directly into the corn genome. The gene produces a crystal protein that is toxic to certain caterpillars, including those of the European corn borer. After the corn borer ingests the protein, the crystal breaks down and releases a toxin that attacks the gut lining. The borers stop feeding within a few hours and die within a couple of days. The presence of this *Bt* toxin in corn provides season-long protection against European corn borers. *Bt* corn offers an opportunity to control corn borer without the use of conventional insecticides.

However, if corn hybrids containing a *Bt* protein are planted widely, corn borer populations eventually will develop resistance to *Bt*. Consequently, producers who grow *Bt* corn should implement a resistance managment plan to slow down the potential onset of resistance. In theory, high doses of the *Bt* toxin in *Bt* corn will kill virutally 100% of the corn borers. However, if any borers survive in *Bt* corn, growers should ensure that surviving adults will mate with adults from areas in which the borers are still susceptible to the *Bt* toxin. "Refuges" where corn borers are not exposed to the *Bt* toxin may be the most practical resistance management tactic. Refuges include all fields of non-*Bt* corn and the more than 200 species of plants on which corn borers can develop. Refuge fields should be adjacent to fields of *Bt* corn, or refuges of non-*Bt* corn should be planted within a field of *Bt* corn. Other resistance management tactics may be developed as more *Bt* corn hybrids become available.

Selected References. 118, 126, 154, 165, 182, 245

Dennis D. Calvin and John W. Van Duyn

European Earwig

Scientific Classification. *Forficula auricularia*, Dermaptera (Forficulidae).

Origin and Distribution. The European earwig was introduced into the United States in the early 1900s. It has spread throughout most of the United States and parts of Canada.

European earwig (Dennis J. Schotzko).

Description. Earwig adults are 5/8 in (16 mm) long, slightly flattened, and red-brown with yellow legs. The forewings are short and leathery, and the hind wings are folded underneath. The tips of the hind wings protrude slightly from beneath the forewings. The tip of the abdomen bears a pair of strong forcepslike structures. The nymphs resemble small adults without wings.

Pest Status. Earwigs are minor pests of corn, generally feeding only in the rows at the edge of a field. Injury is sporadic and is limited mostly to the Pacific Northwest.

Injury. Earwigs injure corn by feeding on silks before pollination is complete. Injury is very similar to that caused by adult corn rootworms. Earwigs also may feed in the tips of the ears, but the injury is minor. Crop injury is most prevalent when weeds or grasses along the field edge become dry early in the season.

Life History. Earwigs complete 1 generation per year in North America. Adults overwinter in field litter or in other protected places. Sometimes eggs deposited in the fall also overwinter. Females deposit most of their eggs in one batch in the spring, occasionally producing a second batch. Eggs hatch over a period of 20 to more than 70 days.

Both nymphs and adults feed on plant material. Although they prefer decaying organic matter, they feed readily on healthy crops or on fruit. Earwigs are active at night and can be found in protected places during the day. They are poor fliers but can fly from a higher to a lower location.

Management. Control measures generally are not necessary. If injury occurs early enough to prevent pollination, an insecticide applied to the edge rows prevents economic damage.

Selected References. 85, 219

Robert L. Stoltz

False Chinch Bug

Scientific Classification. *Nysius raphanus*, Heteroptera (Lygaeidae).

Origin and Distribution. *Nysius raphanus*, the most damaging native North American species of *Nysius*, is the official "false chinch bug." Found from Virginia to Florida west to California and Mexico, it is replaced by *N. niger* (distinct from the European *N. ericae*) in New England, the upper Midwest, the Pacific Northwest, Canada, and Alaska. Mistaken identifications of *N. raphanus* as *N. ericae* complicate interpretation of earlier reports.

Description. Nymphs are ash-gray with brown-white mottling on the back and red mottling on the abdomen.

False chinch bug (Marlin E. Rice).

Adults are 1/8 in (3–4 mm) long, dirty gray, with brown or black markings. *Nysius niger* is similar but slightly larger.

Pest Status. These unpleasant-smelling bugs are generalists that prefer to feed on plants in the mustard and beet families; they rarely are pests of corn. Outbreaks are associated with drought, causing false chinch bugs to abandon spring hosts in early to midsummer and move in hordes into no-till corn and soybeans, other field or vegetable crops, and even into residential areas in search of succulent vegetation.

Injury. Both species of *Nysius* cause similar injury. *Nysius* are gregarious sap feeders, occurring in high densities (70 per square inch [6.25 square cm] on some plants) in localized or large areas of fields. Extensive sap loss and toxic saliva cause wilting and death of leaf tips or whole plants.

Life History. All stages apparently overwinter. Pink-white eggs, deposited in soil and between clustered plant parts, hatch in 4–8 days. Nymphs feed primarily on weeds and develop in about 21 days. Adults live 2–4 weeks. There are 4–6 generations per year.

Management. Drought probably is essential for large-scale crop invasion. No-till or irrigated crops planted near favored spring hosts (canola, cruciferous weeds, chickweed) are at greatest risk. An insecticide sprayed with 20–40 gallons (84.6–169 liters) of water at the bases of the plants may provide control. Reinvasion of crops from untreated weedy areas, however, makes control difficult.

Selected References. 12, 17, 128, 174

Catherine E. Eastman

Fire Ant (Red Imported)

Scientific Classification. *Solenopsis invicta*, Hymenoptera (Formicidae).

Origin and Distribution. The red imported fire ant was introduced into the southern United States from South America in the 1930s. This species spread from the port of Mobile, Alabama, and now is found throughout the Gulf Coast states. A second species, the black imported fire ant, *Solenopsis richteri*, also was introduced but is less widely distributed than *S. invicta*.

Description. Red imported fire ants produce hills or mounds where the colony resides. Disturbance of mounds results in a rapid defensive response by worker ants which quickly run up vertical surfaces in and around the disturbed sites. Worker ants are red-brown with darker abdomens, ranging in size from 1/16 to 3/16 in (1.5 to 5 mm) long. A stinger is evident at the tip of the abdomen.

Pest Status. Fire ants in cornfields have beneficial and detrimental aspects. Fire ants prey upon several pests of corn, including the sugarcane borer and the corn earworm. Conversely, fire ants feed on corn seeds and seedlings and tend aphids, enhancing aphid problems. Also, worker ants can bite and sting repeatedly, producing severe burning and itching. A white pustule usually forms at the site of the sting within 1–2 days. Fire ant stings imperil human health and occasionally are life threatening.

Injury. Worker ants injure seeds by chewing through the seed coat and removing the embryo. Occasionally, the leaflet and root tissue are injured. Injured seeds fail to germinate and injured seedlings die, resulting in a poor plant stand.

Life History. The red imported fire ant is a social insect. The colony is founded by one or more fertilized queen ants. Eggs hatch in 8–10 days, and larvae develop through four instars before pupating. The first worker ants produced by founding queens develop in 27–37 days. Later in the development of the colony, a queen can lay as many as 200 eggs per day. At 85–90°F (30–32°C), larval development requires 6–10 days. The pupal stage lasts for 7–8 days. Thus, the time from egg to adult is 22–23 days for worker ants, which have a

Red imported fire ants (S. B. Vinson).

life span of 30–180 days, depending upon their size. After 2 1/2–3 years, mature colonies may contain more than 230,000 worker ants.

Colonies become mature when winged male and female ants, called reproductives, are produced. Reproductives reside in the colony until they embark on a mating flight.

About 10% of the worker ants in a fire ant colony periodically leave to forage for food. Upon their return, gathered food is shared among worker ants, the queen(s), winged reproductives, and larvae. Only the last larval instar can consume solid food. Although fire ants are omnivorous, their primary diet consists of insects and other small invertebrates.

Management. In late-season corn, fire ant suppression probably is not warranted because the activities of the ant during that time are largely beneficial. However, protection of seeds and seedlings may be desired in fields with a history of seedling loss. Proper planting ensures rapid germination and keeping furrows closed prevents fire ants' access to the seeds. Some seed treatments prevent injury to seeds and seedlings, and some insecticides applied in the seed furrow during planting also are registered for this use.

Selected References. 75, 76, 155, 250

Bastiaan M. Drees

Flea Beetle (Corn)

Scientific Classification. *Chaetocnema pulicaria*, Coleoptera (Chrysomelidae).

Origin and Distribution. Corn flea beetles are native to the Western Hemisphere and generally occur throughout the eastern portion of the United States. Closely related species of flea beetles are worldwide in distribution.

Description. Corn flea beetles are very small, only 1/16 in (1.8 mm) long. They are shiny black and have enlarged hind legs that enable them to jump when disturbed. Eggs are yellow-white and semitranslucent. Full-grown larvae are 1/6 in (4.5–5 mm) long. Most body segments are nonpigmented; only the prothorax and last abdominal segment are somewhat darkened.

Corn flea beetle and leaf injury (Marlin E. Rice).

Pest Status. Although corn flea beetles prefer to feed on corn, they feed on many other host plants, including orchard grass, crabgrass, fall panicum, redtop, witch grass, Kentucky bluegrass, Sudan grass, yellow foxtail, giant foxtail, barley, and wheat. Foxtail, oats, and wheat can sustain corn flea beetles in early spring until corn is available. Sweet corn varieties and sensitive inbred lines planted in seed production fields are more susceptible than commercial corn hybrids to injury caused by corn flea beetles and to Stewart's bacterial wilt, a disease transmitted by the beetles.

Injury. Corn flea beetles injure corn by feeding on the epidermis of leaves, injury that rarely results in economic yield loss. Small feeding streaks or windowpane feeding patches are evidence of feeding by flea beetles. Corn flea beetles also transmit the bacterium *Erwinia stewartii*. This bacterium overwinters both in corn flea beetles and in soil and plant residue. Seedling corn infected by the bacteria may become stunted, wilt, and display linear lesions with wavy edges on leaves. As the disease worsens, yellowing of the leaves moves upward on infected plants. Disease symptoms may occur in certain sweet corn varieties at any stage of corn development; however, the leaf blight phase in dent corn typically does not occur at or after tasseling. The presence of flea beetle feeding scars may be useful in diagnosing the leaf blight phase of this disease.

Life History. Corn flea beetles overwinter as adults at the bases of grasses near cornfields. Beetles emerge in the spring and feed initially on secondary hosts. After mating, females deposit eggs in the soil near the roots of corn plants. The eggs hatch in about 6 days. Larvae complete their development in the soil and begin to pupate in about 14 days. After emergence, adult corn flea beetles feed and mate for the duration of the summer. Beetles become infected with *E. stewartii* by feeding on diseased hosts. As much as 20% of adults that emerge in the spring may be infected with *E. stewartii.* By midsummer, nearly 75% of corn flea beetles may harbor the disease organism.

Management. Mild winters (the sum of the average monthly temperatures for December, January, and February is greater than 90°F [32°C]) improve the overwintering success of corn flea beetles and the bacteria and, consequently, increase the potential for a greater incidence of Stewart's disease. Although most dent corn hybrids are resistant to the wilt phase of *E. stewartii* after they develop beyond the 5-leaf stage, many hybrids remain somewhat susceptible to the leaf blight phase of this disease. If susceptible inbreds are infested with flea beetles before the 5-leaf stage, an insecticide may be warranted when 2–3 beetles per plant are present, and 10% of the plants are silver or white due to the beetles' feeding. Some sweet corn IPM programs in the northeastern United States advocate a treatment when the density of flea beetles is 6 beetles for every 100 plants examined. Early planting dates should be avoided for those inbreds or varieties that are susceptible to Stewart's disease.

Selected References. 1, 69, 82, 173

Michael E. Gray

Grape Colaspis

Grape colaspis adult (Marlin E. Rice).

Scientific Classification. *Colaspis brunnea*, formerly *Colaspis flavida* and *C. beumea*, Coleoptera (Chrysomelidae).

Origin and Distribution. The grape colaspis, a native of North America, is widely distributed east of the Rocky Mountains. It is a pest of corn primarily in the north-central states. Related species, *C. louisianae* and *C. floridana*, are more common in southern states.

Description. The larva is 1/8–1/6 in (3–4 mm) long, slightly curved (comma shaped), and has a plump, white body with a tan head and prothoracic shield, or plate just behind the head. Its three pairs of legs are short. Bunches of hairs arise from bumps on the underside of the abdomen. The tan adult is oval and about 1/6 in (4 mm) long with rows of tiny punctures on its wing covers, making them appear ridged. Colaspis adults sometimes are confused with newly emerged northern corn rootworm adults, which have smooth wing covers.

Pest Status. The grape colaspis is a sporadic pest most often found in corn planted after red clover or mammoth clover, and occasionally in corn planted after sweet clover, alfalfa, or soybeans. Injury is more severe when weather conditions retard the growth of the seedlings.

Injury. The larvae feed on root hairs and may eat narrow strips from the roots. Denuded roots cannot obtain moisture and nutrients efficiently. Injury symptoms above ground include stunting, wilting, purpling of the leaves and stem (indicating a phosphorous deficiency), and browning of the tips and edges of the leaves. Severe infestations may cause plant death and reduced plant populations. Grape colaspis larvae also cause severe injury to soybeans by scarring and girdling the roots.

Feeding by adults on corn silks and the leaves of alfalfa, clovers, grape, soybeans, and strawberries causes no economic damage. They also feed on bull nettle and smartweed.

Life History. The grape colaspis completes only 1 generation per year north of 36°N latitude and 1 complete and 1 partial generation south of 36°N. It passes the winter as a small larva in the soil 8–10 in (20–25 cm) deep. Larvae become active early in the spring, feed on the roots of host plants, and complete their development in May in the southern United States and from mid-June to early July in the Corn Belt. Pupation occurs in an earthen cell 2–3 in (5–7.5 cm) below the soil surface. Adults emerge from the soil during June in the southern states and in July in the Corn Belt. Females lay eggs in the soil near host plants, including patches of smartweed and bull nettle. Eggs hatch in 7–14 days. Newly hatched larvae feed on roots during the latter part of summer and early fall. Nematodes in the genus *Howardula* parasitize grape colaspis larvae, but their impact on colaspis populations is not known.

Management. Planting corn in a field previously planted to red clover is discouraged. Damaging levels of grape colaspis occur infrequently in corn planted after alfalfa or soybeans. No soil-applied insecticide currently is registered for control of grape colaspis larvae in either corn or soybeans.

Selected References. 68, 79, 173

Kevin L. Steffey

Grasshoppers

Scientific Classification. Redlegged grasshopper, *Melanoplus femurrubrum*; twostriped grasshopper, *M. bivittatus*; differential grasshopper, *M. differentialis*; migratory grasshopper, *M. sanguinipes*; and clearwing grasshopper, *Camnula pellucida*, Orthoptera (Acrididae).

Origin and Distribution. All of the pest species of grasshoppers in North America are native. They are distributed widely throughout North America; however, damage seldom is widespread east of the Mississippi River. Although

Redlegged grasshopper (Marlin E. Rice).

Differential grasshopper (Marlin E. Rice).

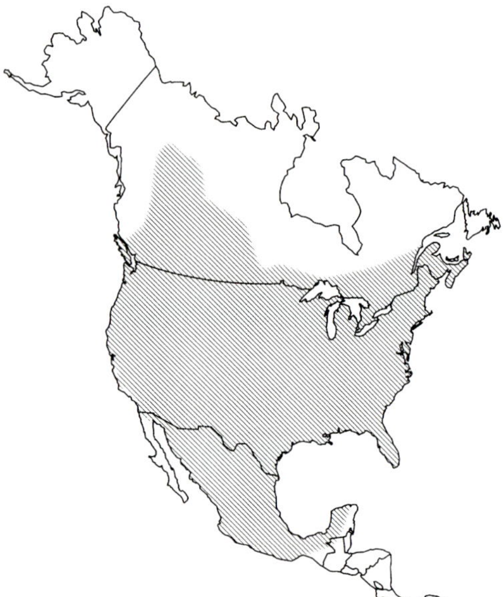

redlegged grasshopper

differential grasshopper

more than 600 species of grasshoppers occur in North America, only a few are serious economic pests. Redlegged and migratory grasshoppers are the most widespread species in North America.

Description. All species of grasshoppers have enlarged hind legs, and the major pests of corn have functional wings. Females of all species are slightly larger than the males. The *Melanoplus* species, the spur-throated grasshoppers, have a "spur" between their front pair of legs. *Camnula pellucida*, a band-winged grasshopper, has clear forewings with dark mottled patches.

The adult redlegged grasshopper is 3/4–1 in (18–28 mm) long and bright yellow on the underside with bright red hind legs. The tan to gray adult migratory grasshopper is 7/8–1 1/8 in (16–29 mm) long; the hind legs are blue-green or red. The yellow-green adult differential grasshopper is 1 1/2–1 3/4 in (22–33 mm) long and has chevronlike stripes on the outside hind femurs. The adult twostriped grasshopper is 1 1/2–1 7/8 in (28–41 mm) long and has a green-tan body with two light yellow stripes along the back of the head to where the wings meet on the back. The adults of the clearwing grasshopper are 3/4–1 1/2 in (20–25 mm) long and yellow to brown.

Pest Status. Grasshoppers infest rangeland and many crops, including corn, alfalfa, clover, cotton, small grains, sugar beets, soybeans, sunflowers, and potatoes. Areas that receive less than 30 in (75 cm) of precipitation in a year are extremely susceptible to grasshopper outbreaks, especially during drought cycles. Injury caused by nymphs usually is more severe at the edges of fields because young, flightless grasshoppers migrate in from field borders. Injury caused by winged, mobile adults is not limited to field edges.

Injury. Adults and nymphs injure corn by feeding on leaf, stem, or reproductive tissues. During outbreaks, entire fields

Twostriped grasshopper (Marlin E. Rice).

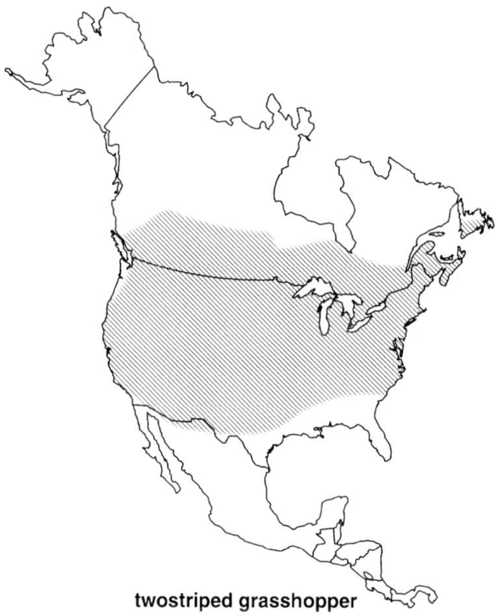

twostriped grasshopper

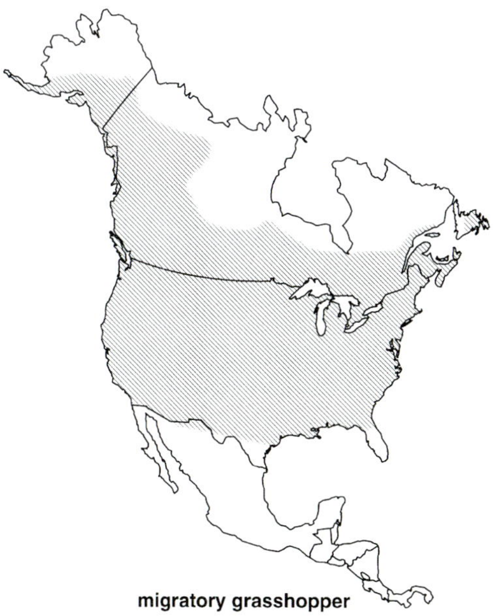

migratory grasshopper

may be denuded. Grasshoppers feed by tearing off and consuming plant tissue with their mandibles, giving the plant a ragged appearance.

Life History. The life histories of all the pest species are similar. The female lays eggs in the soil by forming a burrow with her abdomen. She deposits a cluster of eggs 1/2–2 in (12–50 mm) below the soil surface, then secretes a frothy covering over the eggs; soil particles adhere to the frothy secretion. The secretion hardens, forming a soil-covered egg pod about 1 in (25 mm) long. The number of eggs in each pod varies from less than 10 to more than 100. A single female deposits an average of 200–300 eggs in her lifetime. Most species deposit eggs in uncultivated fields, roadsides, and pastures. These areas are called "egg beds," and the numbers of eggs in the egg beds are extremely large during outbreak years.

Eggs develop in the fall until cold temperatures stop embryological development. Eggs deposited early in the fall are more advanced in development than those deposited later, so they hatch earlier in the spring. Egg hatch can begin in late April but may be as late as early July. A warm spring after a warm fall favors early hatching; a cool fall and spring delay egg hatch.

All pest species of grasshoppers have five (occasionally six) instars. The first instar is very small, about the size of a kernel of wheat, 1/8 in (3 mm). With adequate food and warm, dry weather, nymphal development requires 35–50 days. The species that injure corn complete 1 generation per year and overwinter as eggs.

Management. Weather regulates both the size of grasshopper populations and the effect they have on the crop. Temperatures during the previous summer can influence the current season's grasshopper population. A large number of warm, sunny days from late June through September allows grasshoppers to develop quickly, thus allowing females to deposit large numbers of eggs. Conversely, numerous cool days through the same time period result in fewer eggs being deposited. Extreme, prolonged drought causes eggs to desiccate or stops egg development.

Natural enemies also regulate grasshopper populations. Bee flies, blister beetle larvae, ground beetles, and crickets prey on grasshopper eggs. Predation rates of greater than 50% have been observed in some areas. Spiders, birds, and robber flies prey upon both nymphs and adults. Several pathogens and parasitoids attack both nymphs and adults. The fungus *Entomophthora grylli* causes epidemics in grasshopper populations when the weather is warm and humid. The microsporidium *Nosema locustae* occurs naturally and also is formulated as a microbial insecticide for use in reducing grasshopper populations.

Corn stripped by grasshoppers (Marlin E. Rice).

Grasshoppers feeding on corn (Marlin E. Rice).

Corn ears damaged by grasshoppers (Marlin E. Rice).

Soil tillage has been recommended for several years as a means of destroying grasshopper eggs and exposing them to predators. However, control of grasshoppers by tillage may not be related to egg destruction. Tillage also destroys weeds and volunteer crops required for survival of first instars. Newly hatched nymphs rarely move more than 1 yd (1 m) from their hatch site. If living plants are not available, the nymphs starve. Therefore, tillage is effective for grasshopper control only if all living plants are destroyed before egg hatch. When grasshoppers reach the third instar, they can walk long distances, so tillage no longer is effective. Moldboard plowing or any other tillage operation that buries eggs to a depth that prevents hatching nymphs from emerging from the soil reduces grasshopper densities. However, the risks of soil erosion and reduced soil moisture must be compared with any benefits derived from tillage. Densities of grasshoppers tend to be larger in no-till culture than in conventionally tilled culture.

Specific economic thresholds for grasshoppers attacking corn have not been developed. Control with an insecticide usually is more effective if it is applied during the early stages of grasshopper development. Because grasshopper populations develop outside of the field (e.g., road ditches) early in the season, the insecticide can be applied to a smaller area. Later in the season when adults are present and early season crops such as wheat have matured, corn should be monitored frequently to detect adult migration into the crop. Several insecticides are labeled for control of grasshoppers in corn.

Selected References. 42, 173, 187

Michael J. Weiss

Hop Vine Borer and Potato Stem Borer

Scientific Classification. Hop vine borer, *Hydraecia immanis*, formerly *Hydroecia immanis* (misspelling), *Gortyna immanis*; potato stem borer, *Hydraecia micacea*, formerly *Hydroecia micacea* (misspelling), *Gortyna micacea*, Lepidoptera (Noctuidae).

Origin and Distribution. The hop vine borer, a native of North America, is distributed from Quebec to Virginia and west to the Pacific Ocean, causing the most damage to corn in Illinois, Iowa, Minnesota, and Wisconsin. The potato stem borer, introduced from Europe in the early 1900s, is distributed throughout the Canada maritime provinces south to Massachusetts and west to Manitoba and Wisconsin.

Description. Females deposit brown, pillbox-shaped eggs in small groups or in parallel rows under leaf sheaths or

Hop vine borer (Marlin E. Rice).

Potato stem borer (top) and hop vine borer (bottom) (Bruce Giebink).

dried leaves of grasses. The larvae are 1/12–1 3/4 in (2–48 mm) long, with different color patterns between species. Hop vine borers are dirty white with dark pink or purple transverse bands and four pairs of tubercles, each with a prominent hair, on each body segment. Potato stem borers are similar in appearance to hop vine borers but are red-brown initially, fading to pink as they age. The adults are nondescript, brown or tan, medium-sized "millers" with a wingspan of 1 1/5–1 2/5 in (30–35 mm).

Pest Status. Hop vine borers and potato stem borers cause sporadic, locally severe stand loss in corn. Females prefer to deposit eggs on weedy grasses and occasionally on broadleaf plants, so infestations are associated with weeds. Stand loss typically occurs along grassy field borders, ditch banks, and waterways and terraces. Infestations along field edges usually penetrate less than 20 rows (50 ft, or 15 m) into the field. Poor control of weedy grasses in no-till or conservation-till fields may result in fieldwide infestations. Problems are aggravated by continuous corn production.

Injury. Hop vine borers and potato stem borers tunnel into corn stems below the soil surface. Attacks by hop vine borers and potato stem borers are more lethal than attacks by stalk borers, reflecting both an earlier timing and lower position of attack. Stand loss reduces yield because adjacent, intact plants do not compensate for stand loss.

Life History. The life history of the hop vine borer and potato stem borer is similar to that of the stalk borer, differing slightly in seasonal phenology and host preferences. Each species completes 1 generation per year. Overwintering eggs hatch in late April and early May. Potato stem borers and hop vine borers both have six, occasionally seven, instars.

Newly hatched larvae tunnel into above-ground stems, usually of grasses. Older larvae outgrow the grass stems and crawl to nearby larger-stemmed host plants such as corn. Dispersal usually occurs over several weeks, depending upon stem size and nutritional quality of primary hosts. Hop vine

hop vine borer

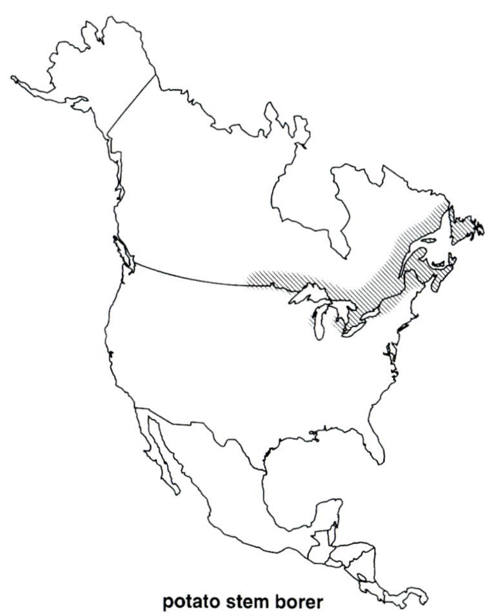

potato stem borer

Hop vine borer tunneling into base of corn stalk (Marlin E. Rice).

borers and potato stem borers disperse primarily as third and fourth instars in late May to early June and mid-June, respectively.

Older larvae feed on secondary hosts. The hop vine borer is rather specialized, preferring only hops and corn. The potato stem borer prefers a broader spectrum of host plants, including potato, tomato, rhubarb, beets, celery, onions, and cereals such as wheat and barley.

Hop vine borers and potato stem borers pupate in early July within soil cells near their last host plant. Adults of both species begin emerging in late July; peak emergence occurs during the first 3 weeks of August. Females preferentially deposit eggs on nearby suitable plants such as smooth brome grass, quackgrass, orchard grass, woolly cupgrass, and wirestem muhly.

Management. Natural enemies and weather contribute dramatically to local fluctuations in abundance. However, persistent infestations in field edges or throughout a field demand management intervention. A simple tactic that diminishes infestations is rotation of corn with nonhost crops such as soybeans or alfalfa. Where crop rotation is not feasible or secondary hosts enable infestations to persist, the management approach differs by type of infestation (field edge or throughout the field) and species. Management begins with the correct identification of the borer responsible and the location of repetitive "hot spots" in the field.

Grassy field edges and grass-covered waterways and terraces are attractive, relatively permanent oviposition sites for these borers. Infestations in field edges can be managed with habitat manipulation or insecticides. Some growers try to eliminate grassy oviposition sites to prevent repetitive, severe infestations. However, eliminating these sites is discouraged because of their value for soil and wildlife conservation. Burning grassy vegetation after oviposition kills eggs but is discouraged because of erosion concerns, fire management hazards, and impacts on wildlife. Mowing before adult emergence can reduce site attractiveness for oviposition. Logistical problems, such as late-season access and wildlife concerns, limit the value of this approach.

Insecticides control infestations along field edges if they are applied just before larvae move into corn. Degree-day information can be used to time an insecticide application. Hop vine borers first begin to move into corn when 500 FDD (278 CDD) have accumulated above a base temperature of 49°F (9.4°C) from January 1; 50% movement occurs when 710–835 FDD (395–465 CDD) have accumulated. Otherwise, an insecticide might be warranted when injury symptoms in corn are first detected. However, both hop vine borers and potato stem borers are difficult to control after symptoms of injury appear.

Infestations throughout fields usually occur where grassy weed problems were evident the preceding fall. The situation is aggravated in conservation-tillage and no-till cropping systems that favor proliferation of some weedy grasses, and in continuous corn where herbicide alternatives are limited. Long-term control depends upon successful weed management and elimination of associated oviposition sites. Early season perennial grasses such as quackgrass also are primary hosts. Applications of "burn-down" herbicides after egg hatch forces the borers into seedling corn. Insecticides should be applied when corn emerges or in combination with herbicides that are specified on the insecticide label.

Selected References. 65, 66, 98, 99

Ken Ostlie

Japanese Beetle

Scientific Classification. *Popillia japonica*, Coleoptera (Scarabaeidae).

Origin and Distribution. The Japanese beetle is native to the main islands of Japan. It was first observed in the United States in 1916 in southern New Jersey and has spread rapidly within the United States since its introduction. It is now found in all states east of the Mississippi River, except in Mississippi and Florida. Isolated infestations also have been observed in Missouri, Iowa, Nebraska, and California (eradicated). The species is established in several southern areas of the Canadian province of Ontario.

Description. The adult Japanese beetle is shiny metallic green with hard, bronze-colored wing covers. Along each side of the abdomen, just below the wing covers, are six tufts of white hair. The adult is about 1/2 in (13 mm) long. The larva or grub is creamy white with a brown head. As the larva grows, fecal material in the hindgut makes the posterior por-

Japanese beetle (Marlin E. Rice).

tion of its body appear gray. The larva is 1/16 in (1.5 mm) long when newly hatched and 1 in (25 mm) long when full grown. The Japanese beetle larva may be confused with other white grubs, such as the masked chafer and May or June beetle. The V-shaped pattern of bristles on the underside of the last abdominal segment, or raster, is characteristic.

Pest Status. The Japanese beetle adult is a sporadic pest of corn. However, it may be a serious pest on other plants, including ornamentals and fruit trees, in areas where corn is grown. It also defoliates soybeans, and soybean fields serve as an infestation source for corn.

Injury. The adult is the main threat to corn, especially during pollination. Beetles may clip a sufficient number of silks to prevent proper pollination. They also feed on leaves; injured leaves appear skeletonized or lacy, similar to injury caused by corn rootworm beetles. However, leaf feeding seldom is economically important.

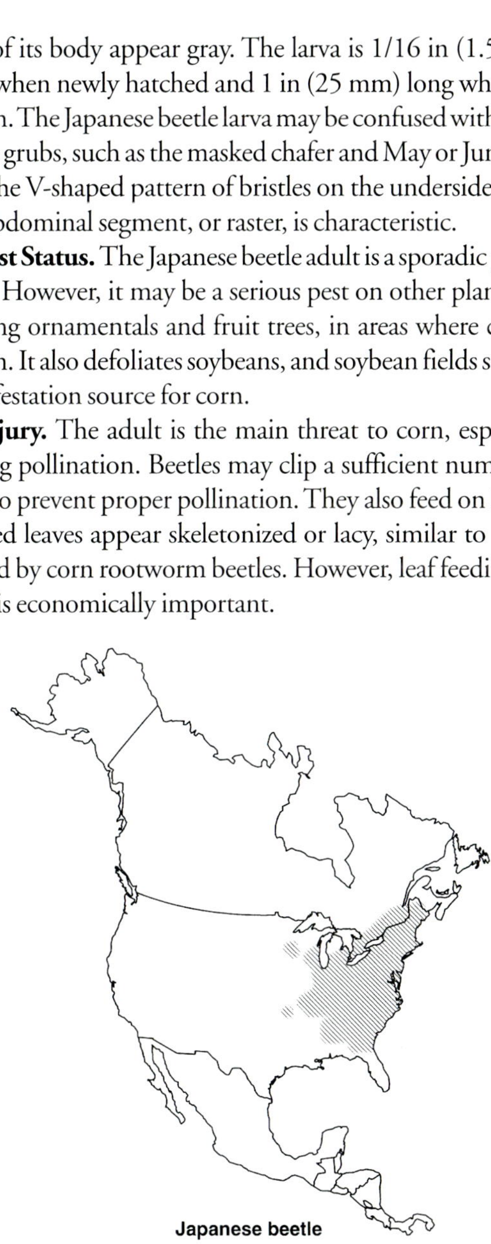

Japanese beetle

Life History. The Japanese beetle completes its development in 1 year in all areas of its range in the United States, except in northern New England, where it completes its development in 2 years. In most areas, the females deposit their eggs in the soil from mid-June through August. The larvae hatch and pass through three instars while feeding on decaying vegetation and grass roots. The nearly full-grown larvae overwinter deep in the soil and move toward the surface in the spring to complete their feeding and to pupate. Pupation takes place in mid- to late May. The adults begin to emerge from the soil in early to late June in most areas of the United States.

Management. Both Japanese beetles and corn rootworm beetles clip corn silks, so the identity of the pest is important. If Japanese beetles are observed in a cornfield, five randomly selected plants in each of five areas of the field should be examined and the number of beetles found on each plant recorded. The length of silks remaining on each plant and the maturity of the silks (i.e., no silks, green silks, some brown silks, or all brown silks) should be estimated. Control with an insecticide may be warranted if silks are clipped to less than 1/2 in (12 mm), fewer than 50% of the plants have been pollinated, and beetles are feeding.

Selected References. 81, 87, 242

C. R. Edwards

Leafhoppers

Scientific Classification. Blackfaced leafhopper, *Graminella nigrifrons*; corn leafhopper, *Dalbulus maidis*, formerly *Baldulus maidis*, Homoptera (Cicadellidae).

Origin and Distribution. The blackfaced leafhopper, a native of North America, is found throughout the southeastern, midwestern, and eastern United States, Canada, and the Caribbean Islands. The corn leafhopper originated in south-central Mexico and has spread to all tropical and subtropical regions of North, Central, and South America and to the Caribbean Islands. In the United States, it is found along Gulf Coast states and in California wherever corn is grown. It does not overwinter in regions where sustained subfreezing temperatures occur.

Description. Immature and adult blackfaced leafhoppers are found on weedy annual and perennial grasses, small grains, and corn. The adult is 1/8–1/6 in (3–4 mm) long. The upper surface of the head, thorax, and forewings are olive green. Four small, round black spots are evident on the anterior margin of the head. Underneath, the background color of the face, thorax, and abdomen is black, mottled with olive green or light brown.

Blackfaced leafhopper (L. R. Nault).

Mexican corn leafhopper (*left*) and corn leafhopper (*right*) (L. R. Nault).

All stages of the corn leafhopper are found in corn, especially in the whorl leaves. Immatures and adults are pale yellow. Adults are 1/8–1/5 in (3–5 mm) long and have a pair of large, round black spots near the margin of the head. A closely related species, the Mexican corn leafhopper, *Dalbulus elimatus*, which has been reported from the southwestern United States, is darker and has two small black spots above and two spots below the anterior margin of the head.

Pest Status and Injury. The blackfaced leafhopper is not found in large numbers on corn and does not cause noticeable feeding injury. However, maize chlorotic dwarf waikavirus (MCDV), which is transmitted by the blackfaced leafhopper, causes serious damage. This virus is found everywhere in the United States where the distribution of this leafhopper and the perennial weed Johnsongrass overlap. In the spring, overwintering or immigrant adult leafhoppers acquire the virus from this weed and spread it to nearby cornfields. Corn leaves infected with MCDV become yellow or red, and the plant may be severely stunted. Highly susceptible corn varieties die prematurely. In regions where MCDV occurs, corn also may be infected with maize dwarf mosaic virus (MDMV). This aphid-transmitted virus also overwinters in Johnsongrass. Symptoms of MDMV are a yellow mosaic or mottling of leaves and plant stunting.

Very large densities of corn leafhoppers occasionally develop when corn is planted sequentially during several months, or when corn is planted year-round and irrigated during the dry season. Withdrawal of plant sap by leafhoppers causes plants to wilt and die. However, damage caused by three stunting pathogens transmitted by the corn leafhopper is even more important than feeding injury. About 3 weeks after corn plants are inoculated with maize bushy stunt mycoplasma (MBSM), leaves turn yellow or red. Plants become stunted, and either no ears are produced, or many small, barren ears develop. Infection by corn stunt spiroplasma (CSS) turns leaves yellow or red; broad, white

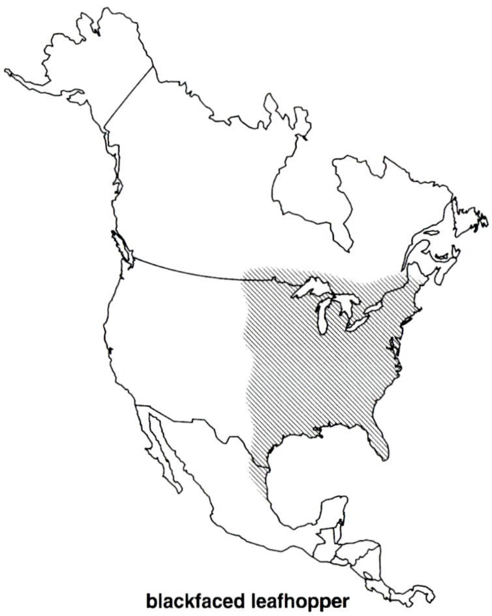

blackfaced leafhopper

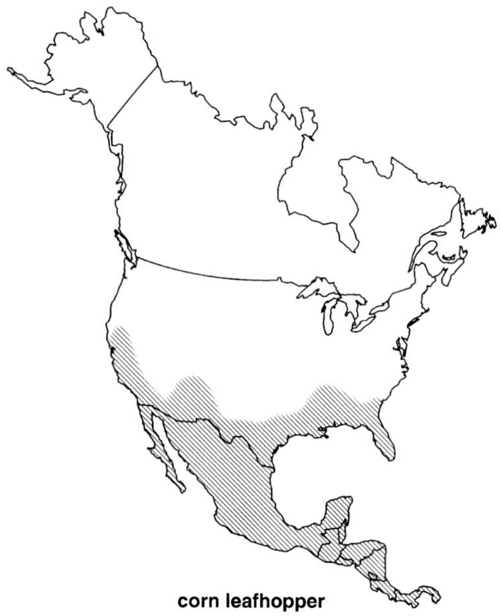

corn leafhopper

longitudinal stripes develop on upper leaves. Severely diseased plants are stunted and do not produce ears. Maize rayado fino virus (MRFV) produces fine white stripes about 1 week after inoculation. Stunting is not as severe as in plants infected with CSS or MBSM.

Life History. Blackfaced leafhoppers can complete development on young corn, but they prefer small grains and weedy annual and perennial grasses. Two to 3 generations develop in northern states, and perhaps twice that many develop in the southeastern United States. Depending on temperature, development from egg to adult requires 3–5 weeks. Blackfaced leafhoppers may develop continually in southern states and overwinter as eggs in perennial grasses or winter grains in northern areas.

In the spring, overwintering corn leafhopper females fly to and settle in the whorls of corn seedlings, where they inoculate the plants with corn-stunting pathogens and lay eggs in leaf tissue. Depending on the environment and corn variety, 2–3 generations develop on the same plant or on nearby corn plants. If corn is planted sequentially, and especially if irrigation is used to grow corn year-round, many generations of corn leafhoppers occur. In regions where corn is not grown year-round, adults disperse (and perhaps migrate) to unknown locations.

Management. Application of granular insecticides at the time of planting reduces the spread of MCDV but is not cost effective. Application of herbicides to control Johnsongrass and planting of MCDV-tolerant corn hybrids are the best management practices.

Foliar insecticides reduce corn leafhopper densities and prevent spread of stunting pathogens, but this practice usually is not cost effective. Year-round production of corn aggravates the problem. A corn-free period is recommended to reduce leafhopper densities. Although strepsipteran and dryinid parasitoids have been identified, they do not regulate corn leafhopper populations. Some corn varieties resistant to corn stunt have been developed.

Selected References. 102, 178, 224

Lowell R. Nault

Leafrollers

Scientific Classification. *Choristoneura parallela* (also known as the spotted fireworm); *Xenostemna pallorana*, Lepidoptera (Tortricidae).

Distribution. *Choristoneura parallela* occurs from Maine to Florida and west to California. *Xenostemna pallorana* occurs from Quebec to Washington and south to Texas.

Description. Mature larvae of *C. parallela* are 1 in (25

Leafrollers (Marlin E. Rice).

Leafroller and injury to corn leaves (Marlin E. Rice).

mm) long with a brown head, olive-green body, and 10 ivory-colored spots on each body segment. Mature larvae of *X. pallorana* are 1 in (25 mm) long with a light-green head and body and lack spots or stripes.

Pest Status. These two insects are rare pests of corn. Injury to small corn plants (4–8 true leaves) has been observed in border rows near grassy areas.

Injury. Larvae roll the youngest two or three leaves together, tie them with silk, and eat the leaf tissue inside the protective webbing. Feeding injury appears as uneven, ragged holes along the edges of the leaves with no particular pattern. Some leaves may be chewed off whereas others remain tied together by the silk. Feeding injury in the whorl does not penetrate to the growing point of the plant.

Life History. Mixed populations of *C. parallela* and *X. pallorana* have been observed in the same cornfield in Iowa during June. Life histories of both species on corn are unknown. *Xenostemna pallorana* larvae also feed on alfalfa, sweet clover, goldenrod, dandelion, and Scotch pine. *Chori-*

stoneura parallela larvae feed on rose, cranberry, citrus, goldenrod, *Vaccinium*, *Phaseolus*, *Gardenia*, and *Myrica*.

Management. Chemical control of leafrollers is not justified because the amount of leaf injury they cause is minor.

Selected References. 134, 158, 199

Marlin E. Rice

Lesser Cornstalk Borer

Lesser cornstalk borer (J. W. Chapin).

Scientific Classification. *Elasmopalpus lignosellus*, Lepidoptera (Pyralidae).

Origin and Distribution. The lesser cornstalk borer is distributed throughout the southern United States from North Carolina to California, north to Washington, D.C., New Jersey, and Massachussetts, and in tropical regions of the Western Hemisphere. It is a pest of corn primarily in the southern states, Mexico, and countries in Central and South America.

Description. Lesser cornstalk borer larvae feed on host plants near the soil surface, in silken tubes attached to a feeding site, or tunnel in the stalk. Full-grown larvae are slender and blue-green with brown bands around each body segment and are 3/4 in (19 mm) long. When disturbed, larvae thrash vigorously. Adults fold their wings when at rest and are about 1/2 in (12 mm) long. The forewings of the female are nearly black; the forewings of the male are light brown, edged with dark brown.

Pest Status. The lesser cornstalk borer is a sporadic pest of corn in the southern United States. Damage often is more severe in dry years and in late planted crops.

Injury. Corn is most vulnerable to attack by lesser cornstalk borers during the seedling stage (2–8 leaves fully emerged). Larvae bore into stalks near the soil surface. Tunneling larvae may destroy the plant's growing tip, located at or slightly below the soil surface during early growth. Larvae tunneling in the center of plants also can destroy part or all of the water-conducting tissues. Symptoms of injury range from stunted and abnormally growing plants to dead or dying plants that appear to be stressed by a lack of moisture. Larvae often move from dying seedlings to adjacent plants, killing several plants in a row.

Life History. The lesser cornstalk borer overwinters as a larva or pupa. Adults emerge in the spring and females deposit eggs on plant leaves or stems or in loose soil near a host plant. Eggs generally hatch in about 7 days. Larvae first feed on leaves and roots, then construct a silken feeding tube from which they bore into plants near the soil surface. Larvae reach maturity in 2–3 weeks, then leave the tube and spin silken cocoons under trash on the soil surface. Adults emerge in 2–3 weeks. Two to 4 generations occur per year in most southern states.

Several natural enemies of lesser cornstalk borers have been reported, but they generally cause little mortality among the borers.

Management. Early planting to avoid infestations of lesser cornstalk borers is recommended in most southern states. Drought conditions and high temperatures favor the development of lesser cornstalk borer populations. Research has shown that lesser cornstalk borer injury to corn is significantly less in no-till plantings than in conventional-till plantings. Producers should consider using a soil insecticide as a preventive treatment when planting conditions (dry weather, late planting) are conducive for infestations of lesser cornstalk borers. Although post-planting insecticide treatments may be effective, adequate sampling methods have not been developed for corn. Generally, by the time infestations are noticed, crop losses cannot be prevented.

Selected References. 173, 244

Robert Lynch

Mormon Cricket

Scientific Classification. *Anabrus simplex*, Orthoptera (Tettigoniidae). The Mormon cricket earned its name by being the plague organism from which the Mormon settlers were saved by the California gull in the Salt Lake Valley in 1848. The Mormom cricket actually is a long-horned grasshopper. Related species, *Anabrus cerciata*, *Anabrus longipes*, *Anabrus spokan*, and the coulee cricket, *Peranabrus scabricollis*, occur within the same range and often have been included in studies and reports about Mormon crickets.

Origin and Distribution. The Mormon cricket is native to western rangeland, particularly sagebrush-dominated

Mormon cricket (Oregon State University).

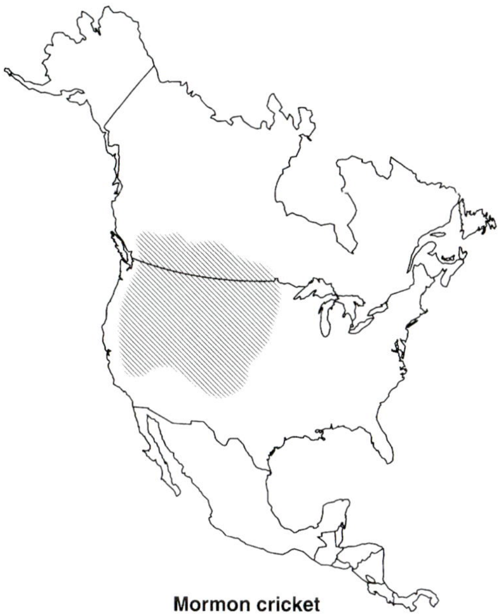

Mormon cricket

habitats. Most pest activity has occurred in Colorado, Utah, Nevada, Oregon, Idaho, Wyoming, and Montana.

Description. The first three nymphal instars are black with white markings; larger nymphs vary from green (at low densities) to black (at high densities). Adults are black, about 1 in (2.5 cm) long, and have small, useless wings and very long antennae. The female is distinguished by a swordlike ovipositor, about equal in length to her body.

Pest Status. The Mormon cricket is primarily a rangeland pest, feeding on more than 250 species of range plants, preferring cultivated crops. Mormon crickets frequently damage corn, alfalfa, and wheat grown in areas where outbreaks occur. They also attack many other crops.

Injury. Injury to corn is similar to that caused by other grasshoppers; the leaves are consumed from the edges inward to the midrib.

Life History. Eggs hatch in early spring, often while snow remains on the ground. The cricket passes through 7 nymphal instars in about 60 days. Mating occurs soon after the nymphs become adults. Eggs usually are deposited singly in light soils on open slopes, where they remain until the following spring.

Young nymphs group together while seeking food and shelter. Later, these small groups merge into migrating bands that can consist of millions of individuals during outbreaks. Bands can travel as many as 50 miles (80 kilometers), feeding on plants in their path, before the onset of oviposition. Band movement then becomes limited to daily forays from favorable oviposition sites to feeding areas.

Many birds, rodents, and mammals feed on Mormon crickets. Cannibalism also is a significant source of natural mortality. Natural enemies include the scelionid egg parasitoid *Sparaison pilosum* and the sphecid wasp *Palmodes laeviventris*. A *Vairimorpha* protozoan is an important pathogen of the Mormon cricket.

Management. The economics of Mormon cricket management on rangeland are poorly understood, but their food preferences probably keep them from competing with cattle under most circumstances. However, when cricket bands encounter corn and other crops, damage usually is significant, so insecticides should be applied before the band enters the field.

Management of Mormon crickets with insecticidal bran baits has been successful; the baits are more selective than broadcast applications of insecticides. *Vairimorpha*-based baits have been employed experimentally. *Nosema locustae*, a protozoan sometimes used for grasshopper suppression, is not effective.

Selected References. 54, 159, 196, 252

Frank B. Peairs and John L. Capinera

Pink Scavenger Caterpillar

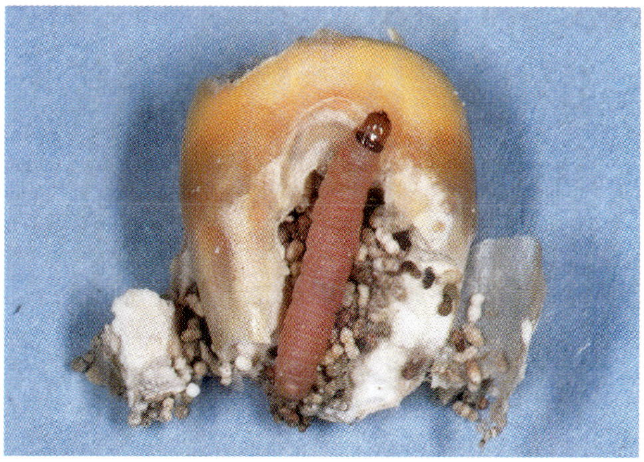

Pink scavenger caterpillar (Marlin E. Rice).

Scientific Classification. *Pyroderces rileyi*, Lepidoptera (Cosmopterigidae).

Origin and Distribution. The pink scavenger caterpillar, formerly known as the pink corn worm, is a tropical and subtropical pest of corn in the southern Atlantic and Gulf states and in Hawaii.

Description. Eggs of the pink scavenger caterpillar resemble Angoumois grain moth eggs. They are almost colorless, flattened, oval, and measure 0.4 by 0.1 mm. Females deposit eggs singly or in groups of as many as four. Newly hatched larvae are white; older larvae are pink to rose and about 2/5 in (10 mm) long. Adults have mottled yellow, brown, and black forewings with long-fringed, narrow hind wings.

Pest Status. The pink scavenger caterpillar is a secondary invader, usually entering the ear after injury has been caused by other insects. Estimated losses have been 2% in the southeastern United States. However, most corn escapes infestations of pink scavenger caterpillars when it is harvested and dried early.

Injury. In susceptible fields, larvae injure the kernels before the hard dough-stage by feeding on the embryo and endosperm. Injured kernels often are webbed. Larvae also may feed on the husk and cob.

Webbing on corn ear caused by pink scavenger caterpillars (Marlin E. Rice).

Life History. The complete life history of the pink scavenger caterpillar is not known. Multiple generations of this pest may occur on an ear when corn is left in the field. Harvesting infested corn carries the insect into storage.

Management. Chemical control is not recommended for this pest. Early plantings of hybrids resistant to primary ear invaders are recommended. Also, early harvesting and mechanical drying prevent additional losses.

Selected References. 49, 68, 168, 233

B. R. Wiseman

Sap Beetles

"Picnic beetles" (*left*) and dusky sap beetle (*right*) (Marlin E. Rice).

Sap beetle larvae and damage to kernels (Marlin E. Rice).

Scientific Classification. Corn sap beetle, *Carpophilus dimidiatus*; dusky sap beetle, *Carpophilus lugubris*; *Glischrochilus quadrisignatus*, sometimes called the "picnic beetle," Coleoptera (Nitidulidae).

Origin and Distribution. The corn sap beetle, dusky sap beetle, and "picnic beetle" are distributed worldwide.

Description. The adult corn sap beetle is about 1/8 in (3 mm) long, ranging from red-tinged black to brown-yellow. The dusky sap beetle is similar in appearance except slightly larger, about 1/6 in (4 mm) long, and dull black with short wing covers and club-shaped antennae. The "picnic beetle" is about 1/3 in (8 mm) long and black with four prominent orange or yellow spots on the wing covers.

Eggs are slender and white. The larvae of all three species are active, white to cream colored with a brown head and posterior, and about 1/4 in (6 mm) long. Larvae are distinguished readily from corn rootworm (*Diabrotica* spp.) immatures by a pair of prominent projections from a dark plate on the tip of the abdomen.

Pest Status. Sap beetles are minor pests of field corn but occasionally serious pests of sweet corn. They usually are associated with injury caused by other insects such as corn earworm and European corn borer.

Injury. Sap beetles usually are secondary invaders of injured plants. Adults have been observed attacking corn borer larvae and forcing them from stalk cavities. Adults may feed on exposed ear tips of corn in the blister or milk stages, but they rarely attack dent corn. If eggs are deposited in injured ears, larvae may cause additional injury to undamaged kernels.

Life History. Sap beetles overwinter in soil and unharvested corn ears as pupae and adults. The adults become active in the spring, feeding on tree sap and decomposing plant material. Females deposit eggs in May and June near or on rotting vegetation, including unharvested corn ears. The larvae complete development in about 30 days and then pupate in the soil. The adults emerge approximately 14 days later and the cycle is repeated. Three to 4 generations each year are common in the central Corn Belt.

Management. Invasion by sap beetles can be reduced or prevented by minimizing injury caused by other insects, especially European corn borer and corn earworm. Direct injury to field corn by sap beetles is usually negligible and noneconomic, so control actions rarely are justified. Sap beetle injury to sweet corn can be reduced by growing varieties with long, tight husks and controlling corn earworms with insecticides. Densities of sap beetles increase in early plantings of sweet corn. Thus, late plantings of sweet corn adjacent to early plantings should be avoided. Sap beetles can be controlled with insecticides. Thresholds have not been developed for these pests. However, control may be warranted in seed production and sweet-corn fields if beetles are causing significant injury to ear tips.

Selected References. 57, 173, 216

L. Von Kaster

Seedcorn Beetles

Scientific Classification. Seedcorn beetle, *Stenolophus lecontei*, formerly *Agonoderus lecontei* and *Agonoderus pallipes*; slender seedcorn beetle, *Clivina impressifrons*, Coleoptera (Carabidae).

Origin and Distribution. Both species are distributed throughout most of North America. The seedcorn beetle is widely distributed in the central and southern United States; a closely related species, *S. comma*, is more common in the northern United States. The slender seedcorn beetle is widely distributed throughout the eastern half of the United States.

Seedcorn beetle (left) and slender seedcorn beetle (right) (Marlin E. Rice).

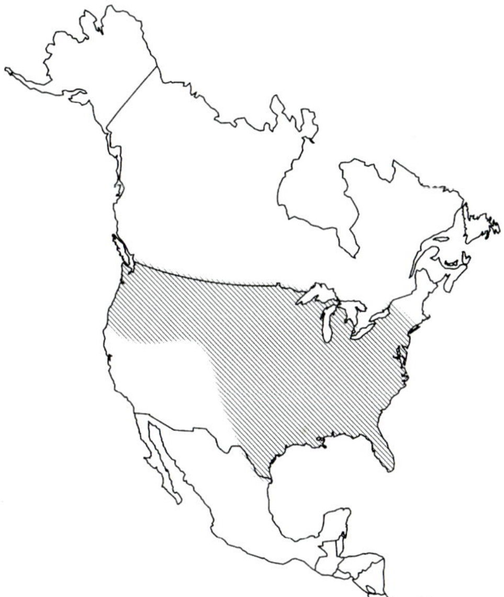

seedcorn beetle

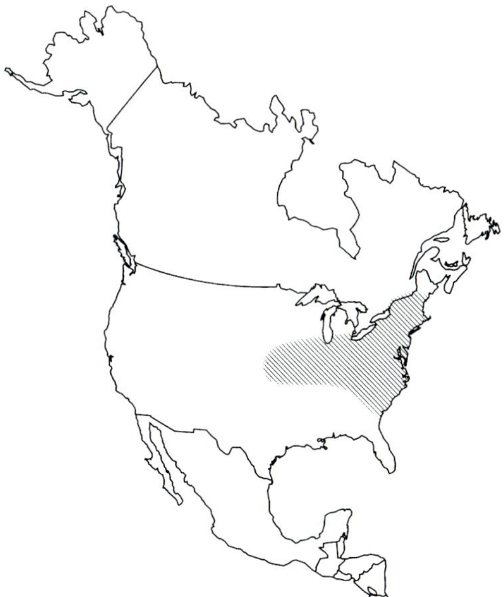

slender seedcorn beetle

Description. The seedcorn beetle is light brown or yellow-brown with a black head and a broad black stripe on each wing cover. Shallow furrows are evident on the wing covers. The slender seedcorn beetle is dark, shiny red, and appears flattened and elongated. The wing covers are wider than the thorax and do not have the black stripes evident on the seedcorn beetle. Both species are about 1/4 in (6 mm) or less in length.

Pest Status. Both species prey on other arthropods, but they also feed on germinating seeds of corn and other cultivated crops when their primary food is absent or in short supply. Damage is more evident during cool, wet weather. Both species are sporadic pests.

Injury. The adult is the damaging stage, attacking the endosperm of the seed. Losses occur because of reduced plant stands. Larvae are predators and do not injure crops.

Life History. Both species overwinter as adults in the soil and complete 2 generations per year. In the central Corn Belt, slender seedcorn beetles are active in the soil as early as mid-March; however, captures in light traps show no evidence of flight of either species until nighttime temperatures reach 68°F (20°C) or more, usually in May. Catches of slender seedcorn beetles in light traps show a single peak in June or July, after which flight has not been observed. Catches of seedcorn beetles in light traps show a first peak in late June or early July and a second peak in September. Larvae likely occur in the soil throughout most of the summer.

Management. Cultural practices are neither useful nor necessary for management of seedcorn beetles. Unusually early planting or planting during prolonged cool, wet conditions may increase the incidence of injury. Planting seed treated with a registered insecticide or applying a registered insecticide at planting to protect corn from damage by rootworm larvae protects the seed from injury by seedcorn beetles.

Selected References. 184, 185

John F. Witkowski and Ramesh Ayyappath

Seedcorn Maggot

Scientific Classification. *Delia platura*, formerly *Hylemia platura*, *H. cilicrura*, *Phorbia fusciceps*, and *Anthomyia zeae* (at least 13 additional synonyms), Diptera (Anthomyiidae).

Origin and Distribution. The seedcorn maggot is distributed worldwide. The related species *Delia florilega* (bean seed maggot) also is widely distributed and has a similar biology and host range.

Description. The adult is gray and smaller and more slender than a house fly (*Musca domestica*) but appears to have longer legs. Unlike house flies, adult seedcorn maggots hold their wings over the abdomen at rest, giving them an oval appearance when viewed from the top. Distinguishing between adults of this species and related species is extremely difficult and can be accomplished only by examining the setae (hairs) on the back and legs. The larva or maggot is the injurious stage. It is smaller than but looks similar to a house fly maggot. The maggot is creamy white, tapered at the front end, and has no legs or visible head. It usually is found in seeds in the ground.

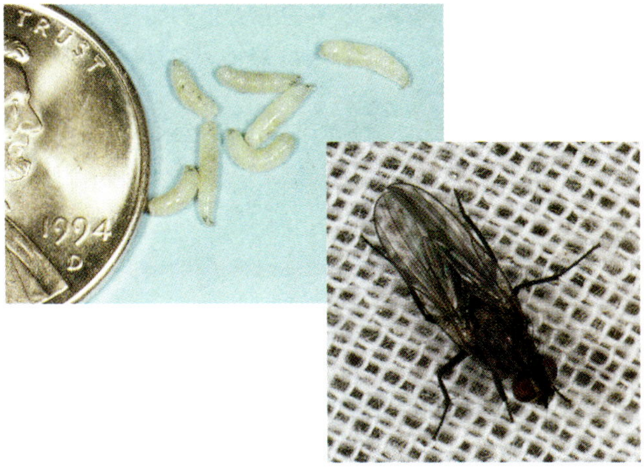

Seedcorn maggots and seedcorn maggot fly (Marlin E. Rice).

Pest Status. Although sporadic in occurrence, this pest can cause serious damage during periods of slow germination in fields with an abundance of green decaying organic matter (especially from legumes) in the soil and in fields that have been heavily manured. Seedcorn maggots rarely are a problem in conservation-tillage fields with soybean or corn residue. They are sporadic pests of at least 50 different crops.

Injury. Maggots tunnel through the corn seed coat causing either a decrease in vigor of the emerging seedlings or complete destruction. Symptoms of leaf injury seldom occur in corn seedlings. Examination of the seed is the best way to detect maggots feeding in corn.

Life History. Seedcorn maggots complete 4–5 generations per year in northern latitudes (about 43°N). In middle latitudes (43–36°N), 2–3 generations occur, with a complete generation in early spring and dormancy of the second or third generation through the summer, fall, and winter. In the far south (36°N and farther south), 3–4 generations may occur between fall and spring. In all warmer regions, only 1–2 generations occur early in the spring because the seedcorn maggot does not complete its development above 75° F (24°C).

The seedcorn maggot overwinters as a brown puparium in the soil. Flies emerge in the spring when soil temperatures increase. A minimum of 65 FDD (36 CDD) is required between adult emergence and egg laying. Females prefer to deposit eggs in recently plowed fields with decaying organic matter (e.g., where manure or green plant residue has been incorporated). Maggots emerge a few days later and attack germinating seeds immediately or feed on decaying organic matter before they move to germinating seeds.

Natural enemies include a cynipid wasp and two species of staphylinid beetles, at least one species of nematode, and some fungi. The most important fungus is *Entomophthora muscae.*

Management. Shallow seeding in well-prepared, warm soil reduces the potential for maggot attack because corn is most vulnerable during seed germination. Planting in cover-cropped fields should be avoided until the organic matter has decomposed, which may take 3 or more weeks. Delaying planting after tillage to allow seedcorn maggots to develop through to pupation has been used as a non-insecticidal management tactic. Seed or furrow treatments with insecticides are effective.

Selected References. 108, 173, 198, 243, 249

C. J. Eckenrode and D. R. Webb

Slugs

Gray garden slug (Marlin E. Rice).

Scientific Classification. Gray garden slug, *Deroceras reticulatum*, formerly *Agriolimax reticulatus*, Mollusca, Gastropoda (Limacidae).

Origin and Distribution. The gray garden slug is an introduced pest from Europe and is commonly found in corn where surface residue accumulates from a lack of tillage.

Description. The adult gray garden slug is 1–2 in (25–50 mm) long when fully extended. It varies from gray to pale cream and exhibits a light mottled pattern of grooves, spots,

Slug injury to corn (Marlin E. Rice).

and streaks. Other slugs associated with no-till corn include the marsh slug, *D. laeve*; banded slug, *Arion fasciatus*; and dusky slug, *A. subfuscus*. The marsh slug is shiny and dark and is about half the size of the gray garden slug. The banded slug is similar in size to the gray garden slug and is light to medium gray with distinct black bands along the sides. The dusky slug is a large orange or brown slug with vertical dark lines on its foot fringe and often dark bands along the sides. An indicator of slug activity on corn is the presence of slime trails (the residue of a viscid substance secreted by the foot and body of a slug as it moves) on the foliage and soil surface.

Pest Status. Slugs are sporadic pests of corn associated with conservation tillage practices, especially no-till. Population densities increase when residues are abundant on the undisturbed soil surface. Crop injury is most severe in the spring when growing conditions are cool and wet. Planting into wet soils usually results in seed furrows that may not close completely, which increases the potential for slug injury to seeds and seedlings.

Injury. Slugs feed on germinating seeds, seedlings, and foliage. Severe stand losses occur when slugs destroy seeds and seedlings in the furrow below the surface residue. Heavy defoliation of young whorl-stage corn plants may retard plant development and cause stand loss.

Life History. Females deposit eggs in masses in the soil; the eggs are held together by a sticky substance that turns yellow before egg hatch. The eggs hatch in about 1 month into small slugs that resemble the adults, except in size. Slugs overwinter as eggs and adults. Survival of the adults is dependent upon the severity of winter conditions. Overlapping generations occur, depending upon the suitability of summer breeding conditions.

Management. Because injurious slug densities are associated with the abundance of residues, tillage that reduces the amount of residue and disturbs the surface habitat minimizes slug problems. Use of row cleaners in no-till fields removes residue above the seed furrows and reduces slug injury to seeds and seedlings. If pre-whorl defoliation is severe or replanting is required, application of a molluscicide bait may be warranted. The bait should be applied to coincide with peak slug activity, which occurs above ground at night when conditions are cool and wet. Rotation of corn with other crops reduces the densities of slugs.

Selected References. 100, 107, 109, 110, 258

Hal Willson and Ronald B. Hammond

Southern Corn Leaf Beetle

Southern corn leaf beetle (Marlin E. Rice).

Scientific Classification. *Myochrous denticollis*, formerly *Colaspis denticollis*, Coleoptera (Chrysomelidae).

Origin and Distribution. In North America, the southern corn leaf beetle occurs from Arizona to Florida in the southeastern United States, throughout the Corn Belt south of Wisconsin and Michigan, and as far east as Virginia and Maryland.

Description. Adult southern corn leaf beetles are 3/16 in (5 mm) long, dark brown, and often covered with bits of soil, rendering them difficult to find in the field. The prothoracic shield just behind the head has three "teeth" on each lateral edge. The larva can be found in the soil and somewhat resembles a grape colaspis larva. It is creamy white and about 1/4–5/16 in (6–8 mm) long. The undersides of the second through eighth abdominal segments each bear a pair of ambulatory processes.

Pest Status. Although reports about this beetle injuring corn date almost entirely from the early 1900s, the southern corn leaf beetle was observed causing injury to corn in Illinois in the 1990s. In some instances, the injury was severe enough that insecticides were applied to prevent additional stand loss, or the plant population was reduced so significantly that replanting was necessary. However, throughout the United States, this beetle has been reported infrequently.

The southern corn leaf beetle occurs most frequently in fields previously devoted to pasture or in fields that have not been cultivated for several years. The beetle also is prevalent in fields infested with cocklebur, apparently another host.

Injury. The adults feed on the stems and chew out notches on the edges of leaves of corn seedlings; injured plants appear ragged. Sometimes the beetles feed in such large numbers that injured plants die. The extent of injury to corn roots caused by the larvae has not been determined.

Life History. The southern corn leaf beetle overwinters as an adult under debris and in clumps of some weed species. Adults emerge early in the spring to feed on young weed hosts, especially cocklebur, and early planted corn. After mating, the female deposits eggs in clusters of 10–50 in weed debris or in soil near corn plants. Eggs hatch in 6–10 days, and the larvae apparently feed on corn roots for about 10 weeks, from early April until mid-June in southern states and from early May until mid-July in the central Corn Belt. The larvae pupate in the soil, and adults emerge from mid-July into August, depending upon latitude. After feeding for a short time, the adults seek overwintering shelter in the late summer or early fall.

Management. Because this insect has been reported so infrequently in corn, economic thresholds have not been established. Preventive insecticides applied at planting time are not recommended. The economic thresholds established for black cutworms could be used as management guidelines. Control of weed hosts, particularly cocklebur, could reduce densities of this insect.

Selected Reference. 138

Kevin L. Steffey

southern corn leaf beetle

Southwestern Corn Borer and Related Stalk Borers

Southwestern corn borer winter (top) and summer (bottom) phases (Marlin E. Rice).

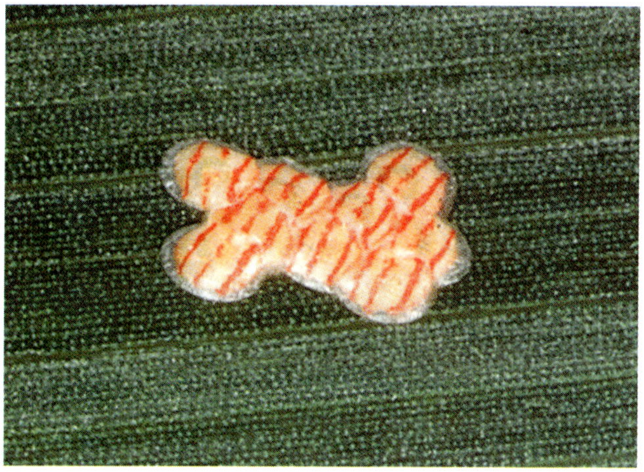

Southwestern corn borer eggs (Marlin E. Rice).

Scientific Classification. Southwestern corn borer, *Diatraea grandiosella*; southern cornstalk borer, *Diatraea crambidoides*; sugarcane borer, *Diatraea saccharalis*; neotropical corn borer, *Diatraea lineolata*; Mexican rice borer, *Eoreuma loftini*, Lepidoptera (Crambidae).

Southwestern Corn Borer

Origin and Distribution. The southwestern corn borer is a native of south-central and northwestern Mexico. This species was first identified in the United States in New Mexico in 1913, although it may have been in the southwestern United States much earlier. From the early 1900s through the 1970s, the southwestern corn borer dispersed across the southern and south-central United States and is now found from Arizona to Alabama north to Kansas, southern Missouri, and Illinois. The northern distribution apparently is limited by the interaction of low temperature, soil type, and moisture, which determines survival of overwintering larvae.

Description. Eggs are oval with a flat base and about 1.3 by 0.8 mm. Eggs are yellow-green when first deposited, but within 36 hours they become white with three broken orange-red lines across each egg. Females deposit eggs in masses of 2–3 on upper and lower surfaces of corn leaves. The eggs slightly overlap, much like fish scales. A female lays about 250 eggs during her 5–7 day life span. Larvae are white with a pattern of large, raised black tubercles on each body segment and are 1–1 1/4 in (25–30 mm) long when full grown. The head of first through third instars is black, whereas the head of older larvae is brown. Overwintering larvae lose their spots and are light yellow-white. Pupae are mahogany brown and are found primarily in stalk tunnels. Adults are dull white or buff colored with a wingspan of 3/4–1 1/2 in (20–40 mm); females are larger than males. The mouthparts extend outward into a characteristic snout. Adults settle during the day on corn plants; mating and egg laying occur at night.

Pest Status. Of the four *Diatraea* species that attack corn in the United States, the southwestern corn borer is the most economically important. A 1976 survey ranked the southwestern corn borer as the fifth most important corn insect pest in the United States and estimated the annual loss as 1% of the nation's production. Damaging infestations occur throughout the insect's range, but losses are especially common in the southern Great Plains region of Kansas, Texas, and New Mexico. Although corn is the primary host, the southwestern corn borer can develop on sorghum, Johnsongrass, and a few closely related grasses.

Economic infestations of first-generation southwestern

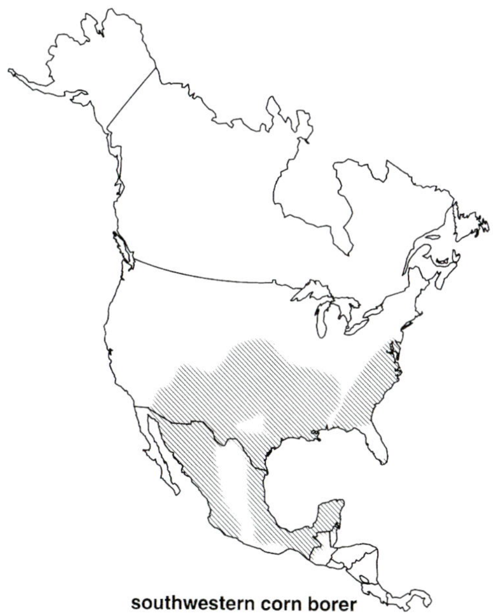

southwestern corn borer

corn borer are uncommon and restricted to corn planted near undisturbed corn stubble fields where adults emerge in the spring. Infestations of second- or third-generation larvae are widely distributed, damage post-tassel corn, and can cause significant yield losses and stalk lodging.

Injury. Southwestern corn borer larvae feed on many corn-plant tissues depending upon plant growth stage. Larvae attacking whorl-stage plants feed on developing leaves in the whorl. Injury is evident as leaves unroll from the whorl. Leaf injury ranges from pinhead sized holes and small circular lesions, or "windows," on leaf surfaces to large, elongated holes. Third instars leave the whorl, crawl down the stalk, and tunnel into the plant, primarily between nodes. Tunneling in small plants may damage the growing point, causing twisted, stunted plants and dead heart. Large larvae bore vertical tunnels in the stalk. Tunneling limits the uptake of water and nutrients by the developing plant and reduces plant height and yield. Yield reductions of 20% and greater have been measured in whorl-stage corn attacked by large densities of larvae.

After females deposit eggs on corn during the reproductive growth stages, newly hatched larvae feed between the husk layers of the primary ear, on ear shoots, and behind leaf sheaths. Larger larvae not only tunnel in the stalk but also may feed in ear shanks and on other ear tissue such as kernels. During late summer and fall, mature larvae crawl or tunnel to the root crown and excavate an overwintering cell in the base of the stalk just below the soil surface. The full-grown larva usually girdles the stalk internally 1–6 in (2–15 cm) above the soil until only a thin epidermal layer remains to support the stalk. At the site of the girdled stalk, the larva seals the tunnel with frass and silk and descends to the crown to overwinter. Winds easily break girdled stalks; fields with many lodged plants just before harvest are dramatic evidence of infestation by southwestern corn borers. Stalk lodging is the primary reason corn producers recognize southwestern corn borer as a serious pest.

Life History. The southwestern corn borer completes 2 or 3 generations annually, 2 in the western Great Plains and 3 in Arkansas, Mississippi, and Alabama. Overwintering larvae pupate in the spring, and adults emerge from corn stubble to lay eggs on whorl-stage corn. Eggs hatch in about 5 days. Larvae feed in corn whorls and later tunnel into the stalk where they pupate. Larvae of the nondiapausing (summer) generation(s) complete 5–6 instars during 18–20 days of development. Larvae of the diapausing generation develop in late July and August and complete six instars during 50–60 days. Mature, pre-diapausing larvae excavate an overwintering cell in the root crown and usually girdle the stalk. Larvae then molt into a nonspotted form that lacks the black

Southwestern corn borer feeding injury in whorl leaves (Allen Knutson).

Southwestern corn borer in base of corn stalk (Allen Knutson).

Stalk girdled by southwestern corn borer (Allen Knutson).

spots characteristic of the summer form, indicating the beginning of diapause. Larvae are cannibalistic. Only one larva overwinters in each root crown.

Important natural enemies of the southwestern corn borer include *Trichogramma* spp. that parasitize the egg; the insidious flower bug, *Orius insidiosus*, and other generalist predators that attack small larvae; and the parasitic nematode *Heterorhabditis* spp. The fungus *Beauveria bassiana* kills many larvae during the fall and winter; epizootics may kill 50% or more of the larvae. Larvae and pupae rarely are attacked by insect parasitoids.

Management. Disking or shredding corn stalks early in the winter greatly reduces survival of overwintering southwestern corn borers and subsequent infestations by first-generation borers. This cultural control practice is the key component of an IPM program for southwestern corn borers. Early planting reduces the threat of heavy infestations later in the season. Stalk lodging is minimized by early harvest, although this practice may be expensive because of the cost of drying corn for storage. Special attachments often are added to combines to gather lodged stalks at harvest. Foliar insecticides are effective if applied soon after egg hatch and before young larvae tunnel into plants. An IPM program in Texas relies on stalk destruction to reduce survival of overwintering larvae and on properly timed insecticide applications for controlling second-generation infestations. Adult emergence is predicted using a degree-day model, and flights are monitored with pheromone traps. The need for and timing of insecticide applications are determined by field scouting by using a treatment threshold of 20–25% of the plants infested with second-generation eggs or with newly hatched larvae. Host plant resistance is being developed by traditional and transgenic means and offers promise in the future. *Bt* corn has been used successfully for managment of southwestern corn borers and offers an opportunity to control this pest without the use of conventional insecticides. Refer to the "European Corn Borer–Management" section (p. 81) for more details about transgenic corn and resistance management strategies.

Southern Cornstalk Borer

The southern cornstalk borer occurs from Maryland and Ohio south to Alabama and northern Florida. The pest has a long history in the southern United States and was first described in 1880 causing serious damage to corn in South Carolina.

The appearance of the life stages, types of injury, and life history of the southern cornstalk borer are similar to those of the southwestern corn borer. The adult is dull white or straw colored and similar in size and appearance to the adult southwestern corn borer. Females deposit flat, oval eggs singly or in masses of 3–15 on corn leaves. Eggs are white when first deposited, but they soon turn orange. Three transverse orange bars on the egg are visible only with the aid of magnification. Eggs hatch in 7–10 days and larvae feed in the whorl or on leaf collars. Larvae are identical to southwestern corn borer larvae and can be distinguished only by examining hair patterns on the body.

Like the southwestern corn borer, southern cornstalk borers feed in the whorl and later tunnel in the stalk. Tunneling is especially extensive in the lower region of the stalk just above the soil line. Reduction in corn grain weight has been estimated to be 6% per tunnel per plant. Heavy infestations weaken stalks so that they may break over. They complete 2 generations annually. Mature, second-generation larvae molt to a nonspotted form that overwinters in a cell excavated in the crown of the plant, much like the southwestern corn borer. However, the southern cornstalk borer does not girdle the stalk in the fall. For this reason, it is not as injurious as the southwestern corn borer, and insecticides rarely are applied to control southern cornstalk borers. Disking cornfields in the fall to expose overwintering larvae to winter weather reduces their survival.

Bt corn has been used successfully for management of southern cornstalk borers. Refer to the "European Corn Borer—Management" section (p. 81) for more details about transgenic corn and resistance management strategies.

Sugarcane Borer

The sugarcane borer entered the United States from the West Indies about 1855 and is found in the sugarcane-growing regions of southern Florida; the Gulf Coast of Texas, Louisiana, and Mississippi; and the Rio Grande Valley of Texas. This borer is a serious pest of sugarcane but also feeds on corn, rice, sorghum, and wheat. Corn is infested only when it is grown near sugarcane because sugarcane stubble and crop residues are overwintering sites for larvae.

The adult is straw colored with two lines of black dashes that create a V-shaped pattern on the forewings. Development from egg to adult requires 30–40 days; 4–5 generations are completed annually in Louisiana. Females deposit flat, oval eggs on host leaves in masses of about 25 each. The creamy white eggs darken to yellow or orange as they develop. Larvae are tan with indistinct brown spots on each segment. Like other *Diatraea* borers, overwintering larvae lack spots.

Larvae feed in whorl- and reproductive-stage plants similar to southwestern corn borers, except sugarcane borer larvae do not girdle the stalk. Yield losses result from reduced ear weight. Infestations during early whorl to early tassel stages of plant development decreased corn yield by 9–14% per borer per plant in studies in Louisiana. Management practices include early planting, not planting corn near sugarcane, and the use of insecticides. Biological control with parasitic wasps has been successful in Texas and Florida.

Mexican Rice Borer

The Mexican rice borer is found throughout much of Mexico and in southern Texas. The borer has been reported from ornamental plantings of sugarcane in southern California and Arizona. This pest causes serious losses in sugarcane and also attacks corn, sorghum, and rice.

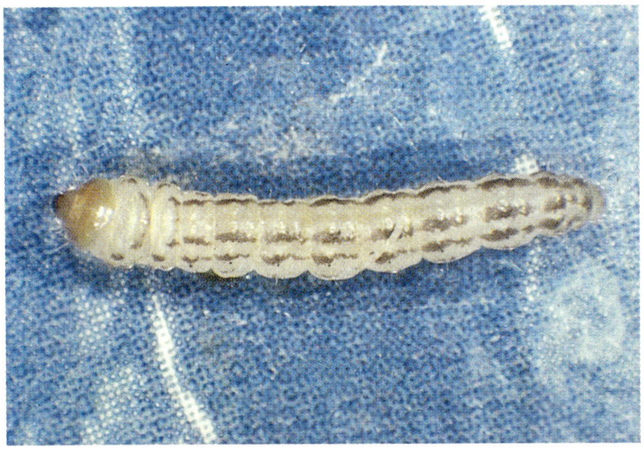

Mexican rice borer (Allen Knutson).

The adult is similar in appearance to the southwestern corn borer but can be distinguished by a small dark spot in the center of each forewing. Females deposit masses of 5–100 cream-colored, globular eggs hidden between layers of dry leaves, making them very difficult to find. Larvae are cream colored with two broken, purple-red lines along the length of the body on each side.

Infestations in corn occur from tasseling to senescence. Young larvae feed in leaf sheaths before tunneling into the stalk. Unlike other stalk borers in the United States, the Mexican rice borer tunnels primarily near the outer surface of the stalk adjacent to the rind, often tunneling both horizontally and vertically. Also, tunnels are packed with frass, unlike those of other *Diatraea* borers that push frass out of their tunnels. Large infestations weaken stalks and cause stalk breakage. Mexican rice borers remain active throughout the year, moving from one host to another.

Injury to corn in Texas is minimized by planting and harvesting early. Many species of parasitic wasps have been imported and released to control the Mexican rice borer.

Neotropical Corn Borer

The neotropical corn borer is found in the United States only in the Rio Grande Valley of Texas, although it is widely distributed in Mexico, South and Central America, and the Caribbean. It is primarily a pest of corn and sorghum but also feeds on related grasses.

The adult and larval stages are similar in size and appearance to those of the southwestern corn borer. Distinguishing larvae of the neotropical corn borer from southwestern corn borer larvae is difficult. However, geographic location of the larvae in the United States distinguishes among species because neither the southwestern corn borer nor southern cornstalk borer occurs in the Rio Grande Valley of Texas. Females deposit flat, oval eggs in groups of two on corn leaves. Eggs initially are white but soon develop three bright red, transverse lines, much like eggs of the southwestern corn borer. Neotropical corn borers feed on and injure corn plants much like the other *Diatraea*. Mature larvae molt to a nonspotted form and enter dormancy in dead corn stalks during the dry season.

Selected References. 34, 47, 89, 127, 140, 175, 182

Allen E. Knutson and Frank M. Davis

Spider Mites

Scientific Classification. Banks grass mite, *Oligonychus pratensis*; twospotted spider mite, *Tetranychus urticae*, Arachnida, Acari (Tetranychidae).

Origin and Distribution. The Banks grass mite is native to western and southern North America and is found as far south as Central America and in Puerto Rico, Hawaii, and Africa. It has more than 80 grass hosts and is a pest of Bermuda grass, bluegrass, corn, dates, sorghum, sugarcane, and wheat. The twospotted spider mite is a worldwide pest of many crops.

Spider mites were first recorded as pests of corn in Texas and California in the early 1950s. Banks grass mites and twospotted spider mites are the most important species, although *Oligonychus stickneyi* and the carmine spider mite, *Tetranychus cinnabarinus*, also attack corn.

Description. Mite species identification is critical because of their different responses to miticides. Characteristics used to distinguish between the two species in the field include the copious webbing produced by twospotted spider mites and the location of the food balls of the adult females. In Banks grass mites, green food balls accumulate around the periphery of the abdomen; in twospotted spider mites, the food balls concentrate in

Twospotted spider mite and eggs (left) and spider mite injury (right) (Marlin E. Rice).

two spots on either side of the abdomen.

Eggs are pearly white and spherical. Larvae have six legs and are light to dark green. Protonymphs and deutonymphs are pale to bright green and have eight legs. Adult males have a deep green, pointed abdomen. Adult females are robust and about 0.45 mm long.

Pest Status. The Banks grass mite is a major pest of irrigated corn produced west of the 100th Meridian. Outbreaks of twospotted spider mites on corn have been distributed more widely in the United States. However, the pest status of twospotted spider mites on corn is less consistent than that of the Banks grass mite. The two species have similar effects on corn yields. Although grain yield reductions in untreated corn in Colorado have averaged about 20% over 25 years, grain yield losses as high as 48% have been reported.

Injury. Spider mites feed by piercing corn leaf tissue and then ingesting liquid cell contents. Feeding causes localized cell damage and eventually a loss of water and chlorophyll and a reduction in dry matter and grain production. Banks grass mite injury begins as concentrated chlorotic areas along the midrib and folded areas of the leaf. Chlorosis can spread, particularly on the basal half of the leaf. Severe damage results in the death of tissue, first of leaf edges and then of the entire leaf. Chlorotic areas resulting from feeding by twospotted spider mites are less concentrated and more evenly distributed across the leaf. Most spider-mite feeding and injury occur between pollen shed and denting.

Life History. Both Banks grass mites and twospotted spider mites overwinter as females, which may deposit some eggs during prolonged warm spells. Life stages of spider mites are: egg; 6-legged larva; 8-legged protonymph; 8-legged deutonymph with robust females and slender males; and adult. Immatures become quiescent before molting. Generation time depends on temperature and is typically 10–20 days. In the laboratory, densities of Banks grass mites can increase 70-fold in 1 generation.

The Banks grass mite overwinters on alternate grass hosts. They return to corn either by walking short distances or by being windborne on silk threads over longer distances. Infestations begin on the undersides of lower leaves and gradually move into the upper part of the plant. The Banks grass mite occurs in corn from the mid-whorl to the grain-filling growth stages. Twospotted spider mites rarely appear in corn before flowering. Rapid population growth most commonly occurs after pollen shed. Factors that encourage infestations of Banks grass mites include host drought stress, elevated temperatures, low rainfall, low humidity, absence of twospotted spider mites, lack of natural enemies, insecticide use, and adequate moisture for alternate hosts during the previous growing season. Most of these factors also encourage buildup of populations of twospotted spider mites, although optimum temperatures for their development (86–90°F, or 30–32°C) are lower than those for Banks grass mites (96–99°F, or 36–37°C).

The distribution of Banks grass mites in corn initially is clumped. Banks grass mites are most abundant in the lower one-third of the plant and density declines as the infestation moves upward on the plant. Twospotted spider mites are more uniformly distributed on the plant.

Management. Management of spider mites in corn currently depends upon scouting, economic thresholds, and miticides. The density of adult females per plant is the best sampling unit, but this is not practical for IPM decision making.

In the past, action thresholds for Banks grass mites were based upon visible injury symptoms in the lower third of the plant, and the presence of mite colonies in the middle third of the plant. A more recently developed threshold based upon percentage infested leaves and the amount of leaf area damaged is used widely. The percentage of infested leaves and percentage of leaf area damaged are compared with information provided in a decision-making table that also includes expected control cost and crop market value per acre. If the field-determined damage is greater than the values provided in the table, a miticide treatment should be cost effective. The economic threshold for twospotted spider mites in corn is similar.

Some miticides are applied preventively and thus have no appropriate action threshold. Because preventive pesticide applications are contrary to IPM philosophy, determining the likelihood of an economic infestation of spider mites before miticide application is recommended. A model available from Kansas State University poses questions regarding the presence of spider mites, natural enemies, field history, alternate hosts, and weather expectations. If responses indicate that conditions are favorable for a spider mite outbreak, a preventive miticide treatment probably is justified.

Miticide resistance is widespread for both twospotted spider mites and Banks grass mites. The level of resistance varies with species, location, and year. Although Banks grass mites generally are less resistant to miticides than twospotted spider mites, they are highly resistant in areas that have long histories of insecticide use in corn and other crops.

Few nonchemical management tactics have been identified for spider mites in corn. However, 35 natural enemy species from 15 families of predatory insects, mites, and spiders have been associated with spider mites in corn. Spider mites also are subject to fungal disease. Augmentative releases of predatory mites have been effective experimentally, but currently are cost prohibitive.

Corn hybrids resistant to spider mites are not available. However, longer-season hybrids develop fewer mite infestations than do hybrids with earlier maturities.

Proper irrigation to avoid drought stress is an important cultural practice for avoiding spider mite outbreaks. However, after mite infestations are established, irrigation cannot reduce spider mite densities in corn. Other recommended cultural practices include late planting and the use of full-season hybrids. Nitrogen fertilization promotes mite infestations, but reducing N application rates to manage spider mites is not economically feasible.

Selected References. 116, 133, 160, 183

Frank B. Peairs and T. L. Archer

Stalk Borer

Scientific Classification. *Papaipema nebris*, formerly *Hydroecia nebris, H. nitela, Gortyna nebris, G. nitela*, Lepidoptera (Noctuidae).

Origin and Distribution. The stalk borer, a native of North America, is widely distributed east of the Rocky Mountains from southern Canada to the Gulf of Mexico.

Description. Females deposit brown, pillbox-shaped eggs in small groups or parallel rows under leaf sheaths or on dried grass leaves. Larvae are 1/12–1 3/4 in (2–48 mm) long, depending upon instar, and purple to black with five longitudinal white stripes (one on top, two on each side) broken by a purple band encircling the body just behind the legs. Adults are nondescript, brown or tan, medium-sized "millers" with wingspan of 1 1/5–1 2/5 in (30–35 mm).

Pest Status. Stalk borers cause sporadic, locally severe stand loss in corn. Females prefer to deposit eggs on grasses and occasionally on broadleaf plants, so infestations are associated with weeds. Stand loss typically occurs along grassy field borders, ditch banks, and waterways and terraces. Infestations along field edges usually penetrate less than 20 rows (50 ft, or 15 m) into the field. Poor control of weedy grasses in no-till or conservation-till fields may result in fieldwide infestations. Problems are aggravated by continuous corn production.

Early instar stalk borer (Marlin E. Rice).

Late instar stalk borer (Marlin E. Rice).

Injury. Stalk borers tunnel into corn stems above the soil surface or climb into the whorl. Young plants (fewer than three leaves) are cut off or wilt rapidly because of the tunneling. In contrast, the center leaves of older plants (4–7-leaf stage) usually discolor, wilt, and die, a symptom called dead heart. After plants grow to the 8-leaf stage, corn usually tolerates feeding with no visible injury symptoms. Nonlethal infestations by stalk borers cause stunting, tillering, delayed development, and increased frequency of barren plants. Whorl feeding by stalk borers results in tattered leaves but has no long-term impact on development or yield.

Life History. Life history of the stalk borer is similar to the life histories of the hop vine borer and potato stem borer, although the stalk borer differs slightly from these species in seasonal phenology and host preferences. The stalk borer completes 1 generation per year. Overwintering eggs hatch in late April and early May. Stalk borers typically have seven to eight instars, with as many as 16 instars reported when nutritional quality of host plants is poor.

Newly hatched larvae tunnel into above ground stems, usually of grasses. Older larvae outgrow the grass stems and crawl to nearby larger-stemmed host plants such as corn. Dispersal usually occurs during several weeks, depending upon stem size and nutritional quality of primary hosts. Stalk borers disperse as fifth to seventh instars in June and early July in the Midwest, but as much as a month earlier in the southern United States.

Older larvae feed on many secondary hosts. The stalk borer has an extremely broad host range that includes more than 176 plant species in 44 families. Stalk borers frequently are associated with broadleaf weeds, particularly giant ragweed and, to a lesser extent, cocklebur, giant burr-elder, docks, and burdock. The stalk borer also is a common pest of tomatoes in vegetable gardens.

Stalk borers pupate in feeding tunnels within host stems or in nearby soil cells in late July to early September. Stalk borer adults begin emerging in late August; peak emergence occurs during the first 2 weeks of September. Females preferentially deposit eggs on nearby suitable grasses such as smooth brome grass, quackgrass, orchard grass, woolly cupgrass, and wirestem muhly and on some broadleaf weeds such as giant ragweed.

Management. Natural enemies and weather contribute dramatically to local fluctuations in abundance. However, persistent infestations in field edges or throughout a field de-

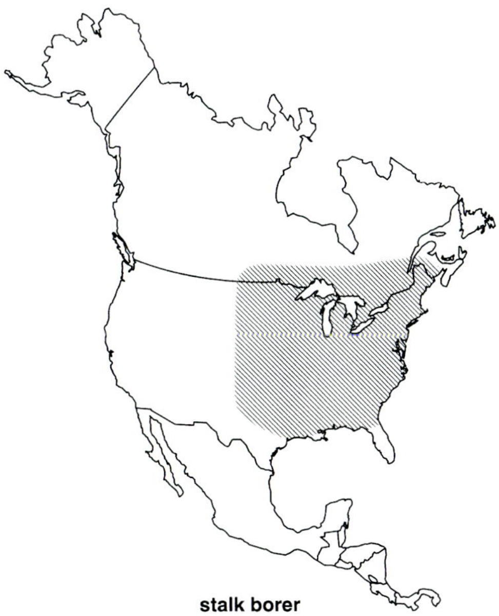

stalk borer

Stalk borer injury to whorl leaves (Marlin E. Rice).

Stalk borer injury to border rows (Marlin E. Rice).

mand management intervention. A simple tactic that diminishes infestations is rotation of corn with nonhost crops such as soybeans or alfalfa. Where crop rotation is not feasible or secondary hosts enable infestations to persist, the management approach differs by type of infestation (field edge or throughout the field). Management begins with correct identification of the borer responsible and location of repetitive "hot spots" in the field.

Grassy field edges and grass-covered conservation features are attractive, relatively permanent oviposition sites for these borers. Infestations in field edges can be managed with habitat manipulation or insecticides. Some growers try to eliminate grassy oviposition sites to prevent repetitive, severe infestations. However, eliminating these sites is discouraged because of their value for soil and wildlife conservation. Burning grassy vegetation after oviposition kills eggs but often is discouraged because of environmental regulations, erosion concerns, fire management hazards, and impacts on wildlife. Mowing before adult emergence also reduces site attractiveness for oviposition. Logistical problems, such as late-season access and wildlife concerns, limit the value of this approach.

Insecticides control infestations along field edges if they are applied to coincide with egg hatch or with movement of larvae from grass hosts to corn. Degree-day information can be used to time an insecticide application. To coincide with egg hatch, an insecticide can be applied when 575–750 FDD (320–417 CDD) have accumulated above a base temperature of 41°F (5°C) since January 1. However, this approach is preventive because the density of stalk borers is not known at the time of application.

Insecticide application also can be timed to coincide with movement of the larvae from grass hosts to corn. Stalk borers first begin to move into corn when about 1,100 FDD (612 CDD) have accumulated above a base temperature of 41°F (5°C) since January 1; 50% movement occurs when about 1,400–1,700 FDD (779–946 CDD) have accumulated. When about 1,300–1,400 FDD (723–779 CDD) have accumulated, corn should be scouted to verify the presence of stalk borers in the grasses (dead stems, larvae inside) or in border rows of corn. Depending upon the leaf stage of the corn plants, insecticide application might be warranted if 15–50% of the plants in the two outside rows are infested. The combinations of leaf stage and percentage infestation are the economic injury levels: 1 leaf (15%), 2 leaf (18%), 3 leaf (23%), 4 leaf (25%), 5 leaf (25%), and 6 leaf (50%).

Infestations throughout fields usually occur where grassy weed problems were evident the previous fall. The situation is aggravated in conservation-tillage and no-till cropping systems that favor proliferation of some weedy grasses, and in continuous corn, where herbicide alternatives are limited. Long-term control depends on successful weed management and elimination of associated oviposition sites. Early season perennial grasses such as quackgrass also are primary hosts. Some insecticides can be tank-mixed with fast-acting "burn-down" herbicides to control the borers that are forced from the dying weeds into seedling corn. If a slow-acting herbicide is used, an insecticide can be applied 7–10 days later.

Some *Bt* corn hybrids offer partial control of stalk borers by suppressing their populations. Refer to the "European Corn Borer–Management" section (p. 81) for more details about transgenic corn and resistance management strategies.

Selected References. 15, 59–61, 147, 149, 182

Ken Ostlie

Stink Bugs (Onespotted and Brown)

Scientific Classification. Onespotted stink bug, *Euschistus variolarius*; brown stink bug, *E. servus*, Heteroptera (Pentatomidae).

Stink bug (Marlin E. Rice).

Origin and Distribution. Onespotted and brown stink bugs are found throughout much of North America but are most common in the Corn Belt.

Description. Onespotted and brown stink bugs are similar in appearance. As adults, both are shield shaped, brown, and measure 2/5–3/5 in (10–15 mm) long and about 1/3 in (7–9 mm) wide. Male onespotted stink bugs have a prominent dark spot near the tip of the underside of the abdomen. Another stink bug similar in appearance to the onespotted and brown stink bugs that is found in corn is *Podisus maculiventris*, the spined soldier bug. However, this species is predaceous. Stink bug nymphs lack fully developed wings and are colorful, displaying orange, yellow, and green against a black background.

Pest Status. Stink bug feeding on corn is sporadic. Potential for injury usually increases when winter weather is mild,

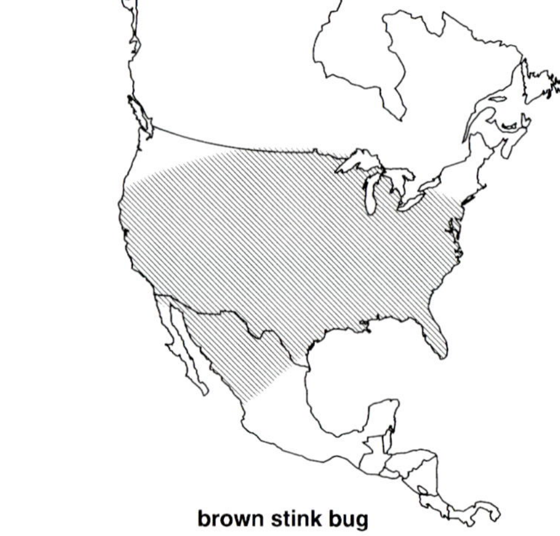

brown stink bug

wheat or rye is planted as a winter cover crop, and corn is planted without tillage.

Injury. Nymphs and adults insert their piercing-sucking mouthparts into the stem and inject digestive enzymes and other compounds that can be phytotoxic or cause growth abnormalities. Symptoms include lines of tiny holes surrounded by yellow to necrotic tissue, twisted leaves or stalks; tillering, stunting, and wilting; and plant death. Yields of injured plants may be reduced by 50% or more. Injury usually is most severe on small plants. As little as 1 day of feeding can result in significant reductions in grain yield. Injury is worst when planters are not adjusted properly and partially open seed slots allow the stink bugs access to the underground stem and growing point of small seedlings. Their feeding on the ear results in aborted kernels, causing the ear to curve outward.

Stink bug damage to corn (Marlin E. Rice).

Life History. Onespotted and brown stink bugs overwinter as adults in vegetation in or near cultivated fields. The most favorable overwintering habitats are alfalfa, wheat, or rye cover crops, or fall-seeded small grains. Stink bugs complete 1–3 generations per year, depending upon latitude. Adults emerge in the spring and feed for 2–4 weeks before females begin laying eggs. Wheat and alfalfa are important spring hosts. Eggs hatch within 3 days, but first instars do not feed. Stink bug nymphs develop through five instars. Older nymphs can be found feeding alongside adults. Stink bugs usually complete their life cycle in 6–8 weeks.

Management. Ensuring closure of seed slots at planting through proper adjustment of corn planters and planting in optimal soil moisture reduce the impact of stink bug feeding. Very few insecticides are labeled for control of stink bugs in commercial corn. Scouting before and after planting is critical. Although, well-defined thresholds have not been developed, thresholds developed for cutworms have been used in some instances. However, if injured plants are considered nonproductive, the relationship between final corn stand densities and yield might be useful for decision making.

Selected References. 9, 77, 84, 217

Marlin K. Bergman

Sugarcane Beetle

Sugarcane beetle (Marlin E. Rice).

Scientific Classification. *Euetheola humilis rugiceps*, Coleoptera (Scarabaeidae). Also called the "rough-headed cornstalk beetle."

Origin and Distribution. The sugarcane beetle is native to North America but is primarily a southern pest; it does not damage corn north of Virginia and Kentucky. The insect is not a pest of corn in the western states; however, it can be found in Central America.

Description. Adults are black "June" beetles about 1/2 in (12 mm) long with a hard-shelled, shiny body. They have

sugarcane beetle

stout legs, and the front pair has stiff spines. The larvae are white, fleshy grubs with a red to brown head, three pairs of well developed brown legs, and a broad brown shield at the tip of the abdomen. Larvae in the soil are curved into a C shape. Depending upon the stage of development, the larvae may be 1/4–1 1/4 in (5–32 mm) long. Pupae are 3/4 in (19 mm) long and are ivory to pale brown.

Pest Status. The sugarcane beetle is a sporadic but serious pest of corn and several other crops, particularly in the seedling stages. Planting corn in fields after grass sod increases the potential for infestation.

Injury. Adults feed on young seedlings at or slightly below the soil surface, producing a large gougelike wound that injures or destroys the growing point and disrupts water-conducting tissues. Plants are killed or growth is impaired. Injured plants are stunted and unproductive. Adults that have overwintered within the field or in nearby grassy areas infest corn within 45 days after planting. The larvae feed on roots of corn later in the season, but they rarely are present in sufficient numbers or at a time when corn is vulnerable to injury caused by their feeding.

Life History. The sugarcane beetle completes 1 generation per year. Adults overwinter in the soil, usually in grassy areas or in fields planted in small grains. They become active in the spring at temperatures that coincide with planting and early growth stages of corn. Severe infestations of sugarcane beetles occasionally destroy entire cornfields planted into grassy areas, especially if minimum tillage or no-till practices are used. The adults mate in the soil soon after becoming active in the spring. Females deposit eggs in groups of 3–4 eggs in small cells excavated in the soil. Females lay 35–40 eggs into June in most locations, and hatch occurs in about 14 days. The larvae feed initially on decaying plant debris, then on the roots of grasses. Average larval development requires about 85 days. Pupation lasts for about 9 days. New adults usually emerge during September and October and are active for a short time before entering diapause.

Management. Avoidance of planting corn into sod or into other grassy areas that have been established for 2 or more years prevents infestations of sugarcane beetles. No-till or minimum tillage practices in grassy areas are discouraged. Early planting and cultural practices that foster rapid and vigorous growth of seedlings enable plants to escape injury and to tolerate large densities of sugarcane beetles in the spring. Recommended insecticides can be used for preventive control. However, the use of insecticides as a rescue treatment is questionable because reliable sampling methods for sugarcane beetles have not been developed.

Selected References. 57, 130, 189

John N. All

Symphylan (Garden)

Garden symphylans (University of Nebraska).

Scientific Classification. *Scutigerella immaculata*, Class Symphyla (Scutigerellidae).

Origin and Distribution. Garden symphylans are distributed throughout North America.

Description. Garden symphylans, also known as garden centipedes, are neither centipedes nor insects. Adults are small and white, measuring 1/4 in (6 mm) long with 12 pairs of legs. Immatures are similar in appearance to adults but have only six pairs of legs.

Pest Status. Garden symphylans are infrequent pests of corn and occur most often in fields with sandy loam soil and a high level of organic matter on which they predominantly feed. Damage often is spotty within a field.

Injury. When densities reach 50–100 individuals per plant, garden symphylans can injure corn early in the growing season by feeding on the root system. Injured plants are stunted and purple.

Life History. Oviposition begins in the spring and continues throughout the summer. Egg masses contain as many as 20 eggs that generally hatch in about 10 days. Garden symphylans are fully developed in 45–60 days, and all life stages occur in the field at the same time. During very cold or hot, dry weather, symphylans move into the subsoil.

Management. If potentially damaging levels of garden symphylans are detected before planting, use of a preplant or planting-time soil insecticide may be warranted. However, garden symphylans should not be confused with other soil organisms that are beneficial.

Selected References. 57, 88

Ronald B. Hammond and Hal Willson

Thrips

Scientific Classification. Corn thrips, *Frankliniella williamsi*, Thysanoptera (Thripidae).

Origin and Distribution. The corn thrips is widely distributed across the southern United States, Mexico, Cuba, Puerto Rico, Hawaii, and the Philippine Islands. It primarily infests corn, sorghum, millet, sugarcane, and many weeds.

Description. Thrips are fragile, small, soft-bodied insects less than 1/12 in (2 mm) long and are pale yellow to white. Adults have two pairs of fringed, featherlike wings. In the northern United States, the pale-bodied *F. occidentalis* and dark-bodied *F. tenuicornis* occur in the whorl, and the pale-bodied *Anaphothrips obscurus* occurs on the undersides of mature leaves.

Thrips (left) and thrips larva (right) (Marlin E. Rice).

Pest Status. The corn thrips is a sporadic pest of corn in the southeastern United States. Heaviest infestations usually are associated with very dry conditions in early spring. As weed hosts mature in the spring, large numbers of corn thrips migrate into seedling corn. The most severe injury has occurred in irrigated corn in drought areas. In the northern United States, thrips also are sporadic pests in irrigated corn.

Injury. Thrips rasp the surface tissue of plants with their mouthparts and suck the exuding plant fluids. Corn thrips prefer to feed on new growth. Their feeding causes a scar or lesion that is a white to gray, 2–3 in (5–7.5 cm) long stripe on the corn leaf, similar in appearance to certain leaf diseases. Thrips injury occurs on leaves, stalks, and husks. Severely injured plants are stunted. Thrips injury at the base of the ear can cause deformed and undeveloped ears of poor quality. Secondary fungal infections of injured ears also have been associated with thrips injury.

Life History. Multiple generations of corn thrips occur annually. Winged adults migrate into nearby cornfields when overwintering weed hosts dry or mature in the spring. Adults feed and mate on corn. Females deposit eggs in small slits that they cut into the corn leaf. Immature, wingless thrips emerge and feed on the corn plant for several weeks. Several hundred thrips of all ages can be found feeding in the whorl, at the leaf base on the stalk, or under the shucks of ears. Thrips molt three to five times as they develop from eggs to adults. A generation may be completed in less than 10 days when conditions are ideal.

Management. The economic importance of this pest on corn in the southeastern United States has not been determined. Examinations of infestations in the field indicate that some yield reduction occurs under some circumstances. Chemical control generally is not warranted because of the large densities of thrips required to cause economic damage, the location of the thrips on the plants, and the difficulty of getting an insecticide to the thrips. Use of planting-time insecticides and management of weedy borders may be the most appropriate means of managing corn thrips.

Selected References. 19, 153

Randy Hudson

Twolined Spittlebug

Spittlebug nymphs (Marlin E. Rice).

Scientific Classification. *Prosapia bicincta*, Homoptera (Cercopidae).

Origin and Distribution. The twolined spittlebug is native to North America and occurs throughout the eastern United States and west into Iowa, Kansas, and Oklahoma. It also occurs in southeastern Canada and probably in northern Mexico.

Description. The adult has sucking mouthparts, is dark brown or black with two red or orange lines across the wings, and has red eyes. Adults are 1/4–3/8 in (6–9 mm) long and resemble large leafhoppers and treehoppers. The nymphs are almost always found in a frothy white mass of "spittle," are usually about the same size as adults, and have ivory-colored bodies, brown heads, and red eyes. Adults are sporadic feeders on all plant parts; nymphs (and spittle masses) are located on brace roots of corn, most often in no-till fields.

Pest Status. The twolined spittlebug commonly occurs in corn throughout the season. Although numerous spittle masses sometimes are present in the surface mulch of no-till fields, spittlebugs are not known to occur in numbers sufficient to cause economic damage.

Injury. Adults and nymphs suck plant juices and cause localized minor injury at the point of feeding.

Life History. The twolined spittlebug occurs commonly in grassy environments. Corn is a host, but spittlebug populations often originate from grassy weeds within or adjacent to cornfields. Nymphs require moist environments, and numerous spittle masses often occur in surface mulches in no-till cornfields. The twolined spittlebug overwinters as eggs in surface debris and under leaf sheaths and hatches in the spring when temperature and humidity are appropriate. The young nymphs immediately produce a frothy spittle mass and develop in this material through four instars for about 1 month under normal conditions. Adults are active and are most easily observed early in the day. Females lay 45–50 eggs during 2–3 weeks. Eggs hatch in about 2 weeks. Spittlebugs complete 2 generations per year throughout much of their range, but probably only 1 generation in northern areas.

Management. The twolined spittlebug does not cause economic damage to corn, so control measures are not necessary.

Selected References. 16, 39, 242

Lowell R. Nault

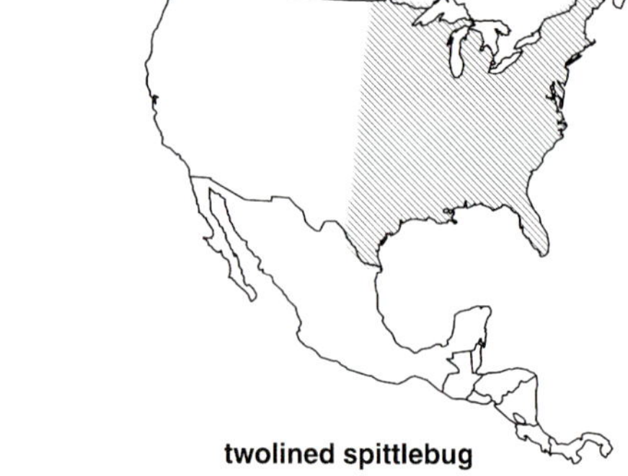

twolined spittlebug

Webworms

Sod webworm (Marlin E. Rice).

Scientific Classification. Numerous species in the genus *Crambus*, Lepidoptera (Crambidae).

Origin and Distribution. Several native species of webworms are found throughout the United States and Canada.

Description. Larvae usually are gray-brown with black spots and coarse hairs. Full-grown larvae are approximately 1 in (2.5 cm) long and very active. Pupation occurs in a silken cocoon a short distance underground. The webbing of the cocoon often contains bits of sand, soil, and plant debris. Adults are pale brown and have a pronounced "snout" formed by the labial palps of the mouthparts. Moths are 1/2 –1 in (1.3–2.5 cm) long and exhibit quick, jerky flight behavior of short duration when they are disturbed.

Pest Status. Webworms are sporadic pests of corn, usually in fields of corn planted into grass sod.

Injury. Larvae defoliate plants, feed on leaf margins, cut plants near ground level, or feed just below the soil surface. Larval tunneling in furled leaves results in straight rows of holes at right angles to the leaf margin, sometimes called shot-holes. The most severe injury occurs when plants are cut below ground or when larvae chew holes into underground portions of the stem and injure the meristem, which results in wilting or profuse tillering of the plant. The silken pupal chambers found underground near the stem distinguish infestations of webworms from infestations by other insects.

Life History. Webworms that attack corn overwinter as larvae inside of cocoons spun the previous summer in sod fields. Sod includes plant species such as native prairie, timothy, annual rye, and bluegrass. Larvae become active in the spring and feed on corn until they are full grown in late June and July. Mature webworm larvae occur as early as June 1 in southern states. Pupation occurs next to corn plants and a short distance underground. Adult emergence in midwestern states begins in late August; flights occur throughout September and early October. Females deposit eggs at the bases of grass plants or drop eggs randomly as they fly over the surface of the sod. Larvae feed for a short time before overwintering in cocoons.

Management. Because injury is confined to first-year corn after sod, the most effective method of control is to avoid planting corn into sod in areas with a history of webworm problems. Plowing the sod before oviposition in late summer also aids in control because webworms do not lay eggs in bare soil. Although no economic thresholds have been developed, thresholds established for cutworms have been suggested for webworms.

Selected References. 3, 91, 127, 173, 255

John L. Wedberg and R. Youngman

Western Bean Cutworm

Western bean cutworm (Marlin E. Rice).

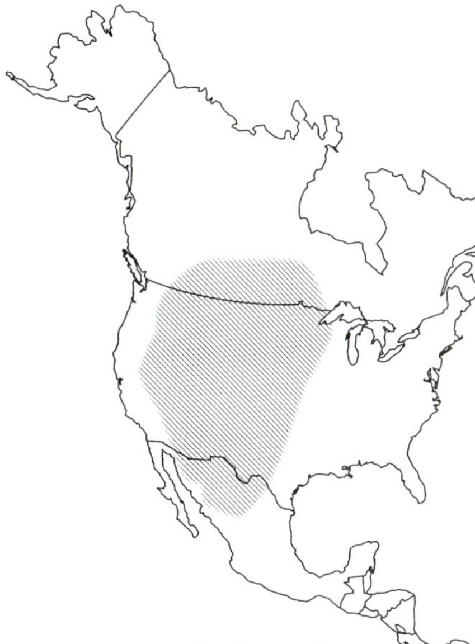

western bean cutworm

Scientific Classification. *Loxagrotis albicosta*, Lepidoptera (Noctuidae).

Origin and Distribution. The western bean cutworm is found from Alberta, Canada, to Mexico and has been reported as far east as Iowa.

Description. Young cutworms are dark brown with faint diamond-shaped markings on their backs. As they grow, the larvae lighten to pale tan, gray-brown, or pink-gray with inconspicuous markings on the body. A mature larva has three short, dark stripes along the top of the first segment behind the head.

Pest Status. Although originally a pest of beans, the western bean cutworm has become a serious pest of corn in some areas of its range.

Injury. Females deposit their eggs on the upper surfaces of corn leaves. After hatching, the larvae move either to the whorl to feed on pollen in the tassel or, if the corn plant has tasseled, to the ear to feed on silks. Pollination may be poor if the cutworms chew off the silks. After the ear has formed, the cutworms feed on developing kernels, destroying as much as 50–60% of the kernels. The loss of kernels, misshapen ears, and entry of fungal pathogens contribute to reductions in corn yields.

Life History. Western bean cutworms complete 1 generation per year. The larvae feed on leaves and pods of green, lima, and pinto beans and on corn. Western bean cutworms overwinter as prepupae in the soil and change to pupae in late spring and early summer. Adults begin to emerge from the soil and deposit eggs in midsummer.

Management. Larvae of the western bean cutworm move from plant to plant, so control with insecticides is recommended if an average of one egg mass or one infested tassel per seven plants is found in the field. Successful control is possible only if at least 95% of the corn plants in a field have tasseled and very few ears are formed before insecticide application. Treatment in Nebraska is suggested if 8% of the plants are infested with eggs or larvae and 95% of the plants have tasseled. After the larvae are inside the husk, they cannot be killed by insecticides and continue to feed until they mature. Application of a recommended insecticide may be necessary to prevent economic losses.

Selected References. 10, 106

Armon J. Keaster

Wheat Curl Mite

Scientific Classification. *Aceria tosichella*, formerly *A. tulipae*, *A. trittici*, *Eriophyes tulipae*, Acari (Eriophyidae).

Wheat curl mites (University of Nebraska).

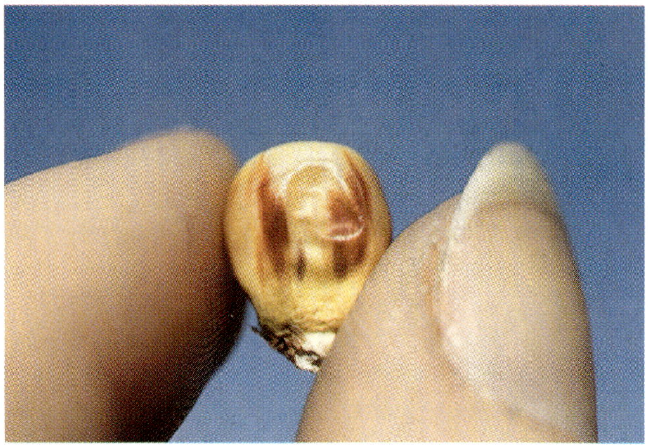

Wheat curl mite damage to corn seed (University of Nebraska).

Origin and Distribution. The wheat curl mite is widely distributed throughout North America, Europe, and Russia and has a broad host range that includes all the cereals and several other grasses. It is primarily a pest of winter wheat in the Great Plains. However, it occurs on corn throughout the Corn Belt.

Description. The wheat curl mite is very small, about 1/100 in (250 μ) long and has only two pairs of legs.

Pest Status. The wheat curl mite transmits the wheat streak mosaic virus and the High Plains disease organism. Wheat streak mosaic is primarily a problem on wheat because most corn hybrids are tolerant or immune to the disease; however, a few inbred lines are susceptible. A field planted with a tolerant corn hybrid may be a source of infestation for an adjacent field of winter wheat. High Plains disease recently was discovered as a problem in corn and wheat in the western Great Plains. Although the disease has caused damage to only a few commercial corn hybrids, many sweet corn varieties are susceptible.

The wheat curl mite also is associated with kernel red streak and corn lethal necrosis. Kernel red streak is common throughout the Corn Belt in some years and is influenced by corn variety and environmental conditions. Corn lethal necrosis is reported only from parts of Kansas and Nebraska and probably results from an interaction of maize chlorotic mottle virus and either maize dwarf mosaic virus (MDMV) or wheat streak mosaic virus.

Injury. The mites first feed on the leaves of corn, and occasionally other hosts, then move to the ears of corn later in the season. Feeding on corn leaves causes little visible injury; feeding on kernels causes red streaking to nearly complete reddening of the pericarp. Kernel red streak does not result in any yield or quality reductions, but the cosmetic injury may be important in white corn. High Plains disease symptoms range from white spotting or a distinct mosaic to elongate, chlorotic spots. Plants infected early may be severely stunted and die. Corn lethal necrosis begins with yellow mottling but progresses to leaf necrosis that extends from the leaf edge inward and from the top downward. Severe yield losses may result.

Life History. The wheat curl mite completes multiple generations in a year, each of which may last 10 days or fewer in warm weather. Mites are distributed among numerous hosts by air movement.

Management. Corn hybrids tolerant to wheat streak mosaic virus should not be planted in areas where this disease is a problem in winter wheat. Varietal tolerance to High Plains disease will be an important management tool for this disease.

Selected References. 8, 27, 132, 180, 224

Gary L. Hein

Whitefringed Beetle

Scientific Classification. *Graphognathus* spp., Coleoptera (Curculionidae).

Origin and Distribution. Whitefringed beetles are distributed widely throughout the southern United States north to New Jersey and west to Missouri and Texas. They are native to South America and were introduced into the United States in 1936, where beetles were found initially on peanuts in Florida.

Description. Whitefringed beetle adults are approximately 2/5–1/2 in (10–12 mm) long and are gray-brown with two white lines along the length of the abdomen. Larvae are creamy white with a light brown head. They are legless and may grow to a length of 1/2 in (13 mm).

Pest Status. Whitefringed beetles are pests of corn, peanuts, tobacco, soybeans, alfalfa, potatoes, cotton, and many vegetables, and are known to infest and feed on at least 385 species of plants. The adults may feed on the leaves, silks, and

Whitefringed beetle (John Matthiessen).

shucks of corn plants; however, their feeding on the foliage of corn and other grass crops does not result in economic damage. Larvae feeding on seedling corn roots and the stem below ground cause the most severe injury. Damaged corn seedlings wilt and often die from the primary feeding injury or from secondary invasions of disease pathogens. Stand losses, reduced vigor of plants, and a higher incidence of lodging caused by the root pruning of the larvae result in yield reductions. Heaviest infestations of whitefringed beetles in corn occur when corn is planted after a legume such as soybeans or alfalfa.

Injury. Whitefringed beetle adults are generalist leaf feeders that only occasionally reach damaging levels. Foliage feeding by adults is distinguished by ragged chewing patterns on the edges of leaves. Larval feeding on plant roots causes the most serious injury. The larvae consume large sections of underground stems and roots. Infested plants exhibit many symptoms associated with root loss. Severely damaged plants often wilt and die, especially if they are stressed by other factors such as dry weather.

whitefringed beetle

Life History. In late summer and early fall, adult whitefringed beetles deposit 200–300 creamy white eggs at the bases of host plants in clusters of 10–15 eggs per plant. The eggs hatch in 10–15 days, depending on temperature. Small larvae feed and develop throughout the summer and the fall on the host plants. The larvae move downward in the soil when temperatures decrease. Although whitefringed beetles overwinter primarily as larvae deep in the soil, they occasionally overwinter as eggs in protected areas. As soon as corn is planted in infested fields, larvae begin feeding on the germinating seeds and developing roots. In late spring, the larvae pupate in the soil for 10–15 days. Adults emerge from the soil in late June and July. All adult whitefringed beetles are female and are incapable of flight. Only 1 generation occurs annually.

Management. Infestations of whitefringed beetles in corn are difficult to control, especially if the infestation is not discovered until late in the growing season. Fields with a history of infestation of whitefringed beetles should be treated with an insecticide applied at planting, especially if corn is planted after a legume crop that was infested with whitefringed beetles. Postemergence applications of insecticides on infested corn are marginally effective. Alternating legume and grass crops in fields with a history of infestation by whitefringed beetles reduces resident populations. Limiting the movement of equipment and crop residue from infested areas into noninfested fields reduces the frequency of infestations.

Selected References. 127, 214, 247

John N. All

White Grubs

Scientific Classification. "Annual" white grub, *Cyclocephala lurida*, formerly *Cyclocephala immaculata*; "true" white grubs, *Phyllophaga* spp., Coleoptera (Scarabaeidae). Adult annual white grubs are called southern masked chafers. Adult true white grubs are called May or June beetles.

Origin and Distribution. *Cyclocephala lurida* is native to North America and found throughout most of the United States, except for the Pacific Northwest. *Phyllophaga* spp. are also native to North America; about 200 species are distributed throughout North America.

Description. The larvae, called white grubs, occur in the soil. They are white and C shaped, measuring up to 1 in (25 mm) long with an orange-brown to brown head, three pairs

White grub (Marlin E. Rice).

of relatively long legs, and a dark posterior. *Cyclocephala* larvae have scattered hairs on the raster; *Phyllophaga* larvae have two parallel rows of stiff hairs on the raster, in addition to scattered hairs.

A true white grub adult is a large (3/4 in [20 mm] long), robust, moderately shiny beetle that is red-brown to black with a darker head and pronotum. The adult annual white grub is stocky and nearly 1/2 in (13 mm) long, with yellow-brown wing covers. The head is dark brown, with a lighter colored "lip" area, giving the appearance that the insect is wearing a mask. The adults are called masked chafers in some states.

Pest Status. Annual white grubs can be serious pests of pasture and bluegrass lawns. Although the annual white grub is not known to cause stand loss in corn, it is often confused with true white grubs that are occasional but potentially destructive pests of corn. Infestations occur most often in fields recently converted from sod. Infestations frequently remain undetected until significant damage occurs.

Injury. Annual white grub larvae may cause minor injury by feeding on seedling corn roots. However, in controlled experiments, densities of nine annual white grub larvae per plant did not significantly injure roots of seedling corn, prevent seedling emergence, or reduce stands.

True white grubs feed on corn roots, often stripping off the root hairs. Corn plants usually reach a height of 2–6 in (5–15 cm) before true white grub injury becomes apparent. Severely injured plants turn light tan or yellow, wilt, and die. Injured plants occasionally turn purple because of their inability to take up phosphorous. Infestations of true white grubs generally are patchy and often located in close proximity to trees, the major food source for adults.

Life History. Annual white grub adults emerge from the soil at dusk during spring and early summer, and females deposit their eggs in the soil. Larvae hatch 2–3 weeks later and feed on grass roots or dead organic matter. During late summer or early fall, larvae move downward in the soil and form earthen cells in which they overwinter. The larvae move back to the soil surface in the spring and pupate in late May or June in Iowa. The adults emerge from the soil in June or July and do not feed. The beetles are attracted to lights at night. As the name implies, annual white grubs complete 1 generation per year.

True white grubs complete a single generation in 3 years in northern areas, sometimes in 2 years in southern areas. Each female deposits 35–60 white eggs in individual cells 1–8 in (2.5–20 cm) deep in the soil. Small, C-shaped, first instars emerge 2–3 weeks after egg deposition and feed on organic matter. Larvae molt once during their first summer. Grubs then move to a depth below the soil frost line. Second instars migrate upward and begin feeding on plant roots in the spring of the second year. Larvae feed heavily on roots the entire second summer; thus, damage usually is greatest in the second year of the life cycle. Grubs molt to the third instar by late fall and again descend in the soil. Larvae ascend in the

June beetle (Marlin E. Rice).

Southern masked chafer (Marlin E. Rice).

spring of the third year and feed until mid- to late summer. Crop stand reduction as a result of feeding by true white grubs also can occur early in the third summer, although damage usually is not as extensive as during the second year. In late July, grubs form earthen cells in which they pupate. During late August and early September, pupae transform to adults, which emerge the following spring.

Recent studies in the northern Corn Belt have shown that densities of *Phyllophaga* are larger in fields bordered by cottonwood or willow trees. The adults feed on cottonwood and willow trees in the surrounding shelterbelts (windbreaks) at night. At sunrise, they drop from the foliage to the soil surface. Females fly from the trees and lay eggs in the soil, usually in those areas that are adjacent to the shelterbelts. Conseqeuently, areas of fields near shelterbelts usually suffer-the greatest damage caused by the white grub larvae.

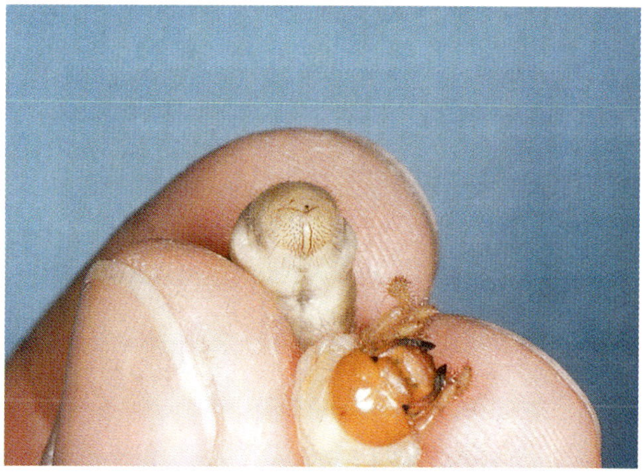

"True" white grub raster (right) (Marlin E. Rice).

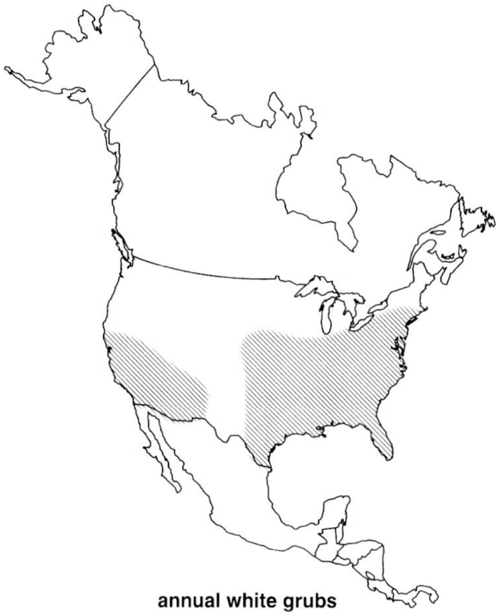

annual white grubs

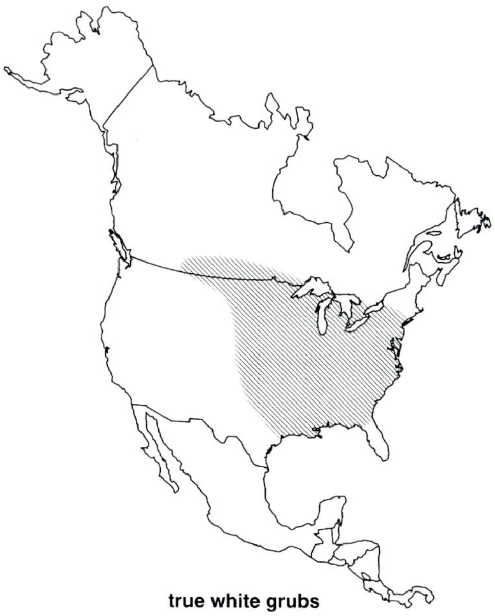

true white grubs

Management. Insecticides are not necessary for control of annual white grubs because they do not injure seedling corn sufficiently to cause economic losses. However, grubs found before planting should be identified correctly to determine whether they are economically damaging true white grubs.

Detection and determination of the level of infestation of true white grubs should be accomplished before planting by taking soil samples in the fall in northern areas or in the spring in southern areas. Samples should be 1 ft by 1 ft by 6 in (30 by 30 by 15 cm) deep. Twenty well spaced samples per 40 acres (16 hectares) are recommended. Densities as low as one larva per square foot may result in significant stand reduction in seedling corn. Several planting-time insecticides are labeled for control of true white grubs. Rescue treatments are not effective.

Selected References. 148, 156, 167, 173, 194, 200, 242

Murdick McLeod, Mike Weiss, and Marlin E. Rice

Wireworms

Scientific Classification. Several species in the genera *Aeolus, Conoderus, Limonius, Hemicrepidius, Agriotes,* and *Melanotus,* Coleoptera (Elateridae).

Origin and Distribution. Wireworms are widespread in distribution and are pests in most of the major corn-growing areas of North America. Wireworm larvae occur at various depths in the soil at all times of the year. Porous, well-drained loam soils are best suited for most species.

Description. Wireworms are cylindrical, wirelike larvae,

Wireworm (Marlin E. Rice).

Wireworm (Marlin E. Rice).

1/3–1 2/5 in (7–35 mm) long, usually hard bodied, and yellow to brown, although some species are soft bodied and white. Adult wireworms are called click beetles because they make an audible click when they flip their bodies into the air, attempting to right themselves when turned upside-down. Adult beetles have hard-shelled bodies that taper at each end. They usually are brown, black, or gray.

Pest Status. Although not considered serious pests in terms of infested acreage, wireworms occur frequently enough to cause severe losses. Large numbers of wireworms often are clustered in relatively small, scattered areas throughout a field.

Injury. Wireworms cause early season injury by boring into seeds before or during germination, preventing germination or killing seedlings before they emerge. Larvae also tunnel into the base of plants below the soil surface, injuring or killing young plants. In heavily populated fields and during the summer months, wireworms bore into the stalks of large corn plants and move up into the plant several inches above the soil. Wireworm injury often is most severe in corn planted after long-standing meadows, pastures, small grains, and other forage crops. Cornfields with chronic infestations left uncontrolled for several years also may suffer heavy damage. Most damage occurs when wireworm populations contain a high percentage of mature or almost-mature larvae.

Life History. Female click beetles deposit their eggs in the soil of grassy areas or cultivated fields. The larvae (wireworms) require 1–5 years to develop into adults. Generations of the long-lived larvae overlap, so wireworms of all sizes and ages can be found in the soil at the same time. However, as the summer progresses and the soil becomes hot and dry, wireworms descend in the soil. They are difficult to find during a dry summer, even in heavily infested fields.

Management. No postemergence insecticides are registered or effective for wireworm control. Therefore, management strategies must be implemented before or at planting time. Wireworms are difficult to detect because their damaging stage occurs below ground. However, a baiting system has been developed to evaluate the potential for a wireworm infestation before planting time. Two bait stations per acre (0.4 hectare) should be established 2–3 weeks before the anticipated planting date, especially in fields where grasses, small-grain stubble, or forage sorghums were present the previous year. Grassy spots in the field and areas where wireworms caused injury last season should be baited. Placement of 5–10 stations per field is practical, but more stations should be established when the field is larger than 25–30 acres (10–12 hectares).

Each bait station should be 2–3 in (5–8 cm) deep and 6–9 in (15–23 cm) wide at the soil surface. One-half cup of an equal mixture of untreated corn and wheat seeds should be placed into a depression that can be filled and slightly mounded with soil. The mound should be covered with an 18-in (46-cm) square piece of black plastic, and this black-plastic sheet covered with a 3-ft (91-cm) square sheet of clear plastic. The edges of the plastic should be covered with soil to prevent wind from blowing away the plastic. The plastic increases the soil temperature, inducing more rapid germination of the bait. A few days before planting, the plastic and soil covering the baits can be removed and the number of wireworms found within and around the bait counted. An average of one wireworm per bait station signifies an economic infestation.

When economically damaging densities of wireworms are present, application of an insecticide at planting time is the best and easiest method for control. Growers also can protect seed from light to moderate numbers of wireworms by adding an insecticide seed treatment to corn seed. However, seed treatment protects only the seed and does not protect the seedling after germination.

Selected References. 173, 177, 201

Armon J. Keaster and Thomas J. Riley

Woollybear (Yellow)

Yellow woollybear (Marlin E. Rice).

Scientific Classification. *Spilosoma virginica,* formerly *Diacrisia virginica,* Lepidoptera (Arctiidae). Adults are called tiger moths. This species also is called the "Virginia ermine moth."

Origin and Distribution. This native insect is a sporadic pest of many crops, including corn, throughout temperate North America.

Description. Woollybear caterpillars get their name from their long, uneven, sparse hairs that produce a fuzzy or woolly appearance. Larvae vary considerably in color and can be creamy white, yellow, tan, gray, or red-brown. Darkening may occur with successive molts. Larvae are 1/4–2 in (6–50 mm) long, depending upon the developmental stage. The pupa is red-brown. Adults have a yellow abdomen, with black spots on the top and sides, and a white thorax from which extend white wings flecked with faint black spots. The wingspan is 1 2/5–2 in (35–50 mm).

Pest Status. Although this insect is found commonly in and around cornfields, it rarely reaches economic levels.

Injury. Larvae feed on leaves and silks. Feeding on leaves produces irregular-shaped notches. Significant silk clipping can occur if flowering is delayed. Injury is most common from midsummer until plant maturity.

Life History. Woollybears complete 1–3 generations per year. Females deposit masses of 100–200 yellow eggs on the under surfaces of leaves in late May and early June. Six instars occur during June and early July. A period of intense wandering occurs during the prepupal phase. An oval cocoon, loosely woven from hair shed during the last molt, is constructed in protected places. Pupation lasts 2–4 weeks. Adults are common from midsummer until fall. Second-generation larvae spend the winter as pupae within their cocoons.

The yellow woollybear is attacked by fly parasitoids and fungal and viral pathogens.

Management. Treatment rarely is justified. No insecticide currently is registered to control yellow woollybears in corn.

Selected References. 67, 162, 173

Larry W. Bledsoe

Beneficial Organisms

Of more than 400 species of insects and mites that occur in a typical cornfield during a growing season, only a small fraction are pests. Insect and mite pests are attacked or infected by many species of beneficial organisms (also called natural enemies), including predators, parasitoids, and pathogens. However, practical information about the use of natural enemies in corn pest management is lacking because beneficial arthropods and pathogens have been studied less intensively than the pest species.

A predator may be an insect or other animal that consumes many insect prey during its lifetime. Predators often are large, active, and/or conspicuous in their behavior. Parasitoids of insects are other insects that lay their eggs in or on the host insect. When the parasitoid egg hatches, the larva feeds on the host and kills it. Many parasitoids are specific to the type of host insect they attack, and they are not harmful to humans. Insect pathogens include viruses, bacteria, fungi, and other microorganisms that cause insect disease. Many insect pathogens attack only one species or a limited group of insects and rarely harm nontarget species.

Classical introduction, augmentation, and conservation of natural enemies are the three primary methods for implementing biological control or enhancing natural control in pest management programs. After a pest has been introduced into the United States from a foreign country, natural enemies of that pest in its country of origin also are introduced into the United States. All of the wasp parasitoids of the European corn borer have been introduced into the United States from Europe and Asia. Although introductions of natural enemies help suppress pest populations, outbreaks of pests still occur. Natural enemies also can be reared and released by humans to augment their natural populations or to inundate a field. *Trichogramma* wasps have been used for this purpose throughout the world, and they show promise for use in some corn production systems in the United States. Conservation of natural enemies encourages their presence. Conservation includes managing cultural practices to allow natural enemies to persist and increase even when the pest is absent.

Some practical uses of natural enemies in corn are discussed in the following sections. Coverage is selective, focusing upon the most geographically widespread, abundant, and easily observed natural enemies of corn pests. The natural enemies of some corn pests remain unknown. Important groups not discussed in this handbook are the Diptera (Cecidomyiidae) that feed on aphids, mites, and other small arthropods; Acari (Phytoseiidae), predatory mites that feed on other mites and some insect eggs; other predatory mites; *Geocoris* spp., Heteroptera (Lygaeidae); paper wasps and hornets that are generalist predators; many parasitoids of aphids; the blister beetle parasites of grasshoppers; and the many families of spiders. The practical significance of most of these natural enemies in corn is poorly understood.

The impact of most species of arthropods in cornfields is not known. Many of these species may be beneficial. Several species of small flies and beetles probably help decompose organic matter by grinding large particles into smaller ones. Some species eat fungal spores, possibly suppressing the incidence of some corn diseases. Ants may recover nutrients from below the rooting zone by bringing soil to the surface as they build their nests. Some minor pests may have a net beneficial effect on corn. Natural enemies may consume minor pests and decomposers and thus survive when key pests are absent. These and many other potentially beneficial organismal interactions in cornfields remain to be characterized.

David Andow

Entomopathogens

Entomogenous Fungi

Scientific Classification. Deuteromycetes and Entomophthorales.

Origin and Distribution. These fungi occur in cornfields throughout the world, causing varying levels of natural mortality among caterpillar, aphid, and mite pests (Table 1). The distribution of these fungi is known primarily from scientific literature from corn-producing areas in Europe, North America, and South America.

Biology. The infective unit of entomogenous fungi is the spore, or conidium. When temperature and moisture conditions are favorable, spores germinate on the host's cuticle to form a germ tube. Optimal conditions for their development are 72–86°F (22–30°C) and greater than 90–95% relative humidity. The tube penetrates by both mechanical pressure and enzymatic digestion of the cuticle. After penetration of the cuticle, the fungus multiplies by budding, spreads throughout the body cavity, and kills the insect within 3–14 days after spores adhere to the cuticle. Death probably results from mechanical damage, nutrient depletion, and the introduction of toxins produced by the fungus. After the host dies, the fungus grows outward through the cuticle and produces spores on the cadaver when environmental conditions are favorable. If conditions are not favorable, the fungus remains inside the cadaver, producing spores when favorable conditions resume.

Grasshopper killed by *Entomophthora* fungus (Marlin E. Rice).

Table 1. Entomogenous fungi attacking corn insects

Host insect	Fungus	Distribution[1]
	Deuteromycetes	
Aphid spp.	*Beauveria bassiana*	NA
	Verticillium lecanii	EU, NA
Corn earworm	*Nomuraea rileyi*	NA
Diatraea spp.	*Beauveria bassiana*	NA, SA
	Hirsutella sp.	NA
	Metarhizium anisopliae	SA
Ostrinia spp. (European and Asian corn borer)	*Beauveria bassiana*	EU, NA, China
Spodoptera spp. (armyworms and cutworms)	*Beauveria bassiana*	NA, SA
	Aspergillus flavus	NA, SA
	Metarhizium anisopliae	NA, SA
	Nomuraea rileyi	NA, SA
	Entomophthorales (Zygomycetes)	
Aphid spp.	*Conidiobolus* spp.	NA
	Entomophthora chromaphidis	NA
	Pandora neoaphidis	EU, NA
	Zoophthora radicans	NA
Grasshopper spp.	*Entomophthora grylli*	NA
Spider mites	*Neozygites tetranychi* (*Tetranychus*)	NA

[1] EU, Europe; NA, North America; SA, South America.

Hosts and Degree of Natural Control. The deuteromycete fungi generally infect caterpillar pests of corn, although several also infect aphids and psyllids (Table 1). Entomophthoralean fungi normally infect aphids, but they also infect mites and cutworms that attack corn (Table 1). The level of natural reduction in populations of corn pests by entomogenous fungi ranges from less than 1% to greater than 85%; the higher and more sustained incidences of infection occur in tropical and subtropical regions.

Biological Control Potential. Entomogenous fungi currently provide only limited natural control of corn pests. In addition, the current marketing environment in the United States does not generate much interest for the

Western corn rootworm killed by *Beauveria* (left). European corn borer killed by *Beauveria* (right) (Marlin E. Rice).

development of candidate fungi for registration as biopesticides for corn. However, some fungi have generated interest in their development as applied biological control agents.

Mass production and application of selected deuteromycete fungi have been used to control pest caterpillars and beetles. *Beauveria bassiana* formulated on granules and applied to corn whorls provided season-long control of European corn borers in experiments in Iowa. Wettable powder, granule, and dust formulations of *B. bassiana* significantly reduced densities of the Asian corn borer, *Ostrinia furnacalis*, in experiments in China. In another experiment in Iowa, injection of a conidial suspension of *B. bassiana* into corn plants significantly reduced infestations by corn borers. The injected fungus colonized the plants and moved within the pith. Commercially formulated *B. bassiana*, *Metarhizium anisopliae*, and other deuteromycete fungi have been tested against corn rootworm larvae (*Diabrotica* spp.) in the soils of cornfields, but results have not been encouraging.

Selected References. 21, 93, 94, 141

Wayne A. Gardner

Viruses

Scientific Classification. Ascoviridae (ascoviruses), Poxviridae (entomopoxviruses), Reoviridae (cytoplasmic polyhedrosis viruses), Parvoviridae (densonucleosis viruses), Iridoviridae (iridescent viruses), and Baculoviridae (nuclear polyhedrosis, granulosis, and nonoccluded baculoviruses).

Biology. Of the seven types of viruses that attack caterpillar pests of corn (Table 2), the granulosis viruses (GVs) and nuclear polyhedrosis viruses (NPVs) are the most common. Both GVs and NPVs have rod-shaped particles containing DNA. NPVs generally infect host larvae when the larvae eat the virus or, less commonly, when a parasitoid injects the virus into a newly encountered host after stinging an infected host. Small or young larvae are more susceptible to infection than are large or older larvae. Infected larvae generally die within 3–10 days. When NPV-infected larvae die, they "melt," releasing more virus that contaminates the plant and can infect other larvae.

Hosts and Degree of Natural Control. Most viruses of insects infect only a single host species or a few closely related species. Thus, they are harmless to humans, farm animals, and nontarget insects such as parasitoids and predators. A few NPVs exhibit wide host ranges, infecting species of caterpillars from more than one family. The NPV isolated from celery looper, *Anagrapha falcifera*, can infect seven corn insect pests (Table 2) and might infect other species that have not been tested.

Most viruses reported from insect hosts in corn are found in caterpillars (Table 2). In addition, a nonoccluded baculovirus and a filamentous virus have been reported from the southern corn rootworm. Also, the aphid lethal paralysis virus and the *Rhopalosiphum padi* virus infect both the bird cherry-oat aphid and the greenbug. A nonoccluded baculovirus has been associated with mortality in *Microplitis croceipes*, a wasp parasitoid of corn earworms. Many more viruses remain to be discovered in insect pests and beneficial arthropods in corn.

Biological Control Potential. The NPVs provide the greatest level of natural control and offer the greatest potential among viruses for applied biological control of insect pests of corn. The broad-spectrum viruses offer more potential for commercial development as biopesticides because of the larger potential market and greater usefulness to the farmer, who usually must contend with more than one species of caterpillar pest.

Selected References. 125, 161, 163

John J. Hamm

Bacteria and Microsporidia

Scientific Classification. Bacterial infections in insects are caused by two distinct groups of bacteria, the nonspore formers and the spore formers. Most spore-forming bacteria are in the family Bacteriaceae, the spore-forming rods.

Several species of protozoa and protozoanlike organisms in the phyla Protozoa and Microspora are pathogens of corn insects (Table 3).

Origin and Distribution. Scientific literature describes nonspore-forming bacteria from many insects. In general, these bacteria are saprophytic in nature and do

Table 2. Viruses[1] reported from caterpillar pests of corn

Pest	AV	EPV	CPV	DNV	IV	GV	NPV	AfNPV
Armyworms								
Armyworm	+	+	+	+	+			
Fall armyworm	+					+	+	+
Yellowstriped armyworm	+							
Corn earworm	+		+		+	+	+	+
Cutworms								
Black cutworm			+			+	+	+
Redbacked cutworm			+			+	+	
Spotted cutworm					+	+		
Variegated cutworm			+			+	+	+
European corn borer						+	+	
Lesser cornstalk borer		+						
Potato stem borer					+	+		
Southwestern corn borer						+	+	
Sugarcane borer			+		+	+		
Yellow woollybear		+			+	+	+	

[1]AV, Ascovirus; EPV, Entomopoxvirus; CPC, Cytoplasmic Polyhedrosis Virus; DNV, Densonucleosis Virus; IV, Iridescent Virus; GV, Granulosis Virus; NPV, Nuclear Polyhedrosis Virus; AfNPV, *Anagrapha falcifera* NPV or celery looper virus. A "+" in a column indicates that the virus has been reported from the pest.

not play a major role in insect management, thus they are not listed in Table 3. The bacteria listed in Table 3 are soilborne bacteria that are worldwide in distribution.

The microsporidia are primarily obligatory, intra-cellular parasites; therefore, their origin and distribution are dependent upon the origin and distribution of their hosts.

Biology. The best-known bacterium in the family Bacteriaceae is *Bacillus thuringiensis*, commonly referred to as *Bt. Bacillus thuringiensis* produces a crystalline protein during sporulation; the protein is toxic to insects. *Bacillus thuringiensis* kills an insect when the immature stage consumes the bacterium with the crystalline protein. The large protein crystal is degraded into biologically active fragments within the gut. These active proteins bind to midgut cells, causing gut paralysis and, subsequently, a septicemia, or an invasion of the bloodstream by the toxins.

Gregarina coronata described from the southern corn rootworm is the only protozoan that infects a corn pest. The spores of this protozoan are ingested by the host insect; sporozoites are released and become imbedded in the cells of the midgut. The gregarine completes its life cycle and passes from the insect in the feces. *Gregarina coronata* is not an acute pathogen; it causes debilitating effects on its host.

Infection of an insect by a microsporidium begins when an immature host insect consumes spores, and the spores germinate, infecting midgut cells. The microsporidium completes its life cycle, sometimes killing the cell. Infective forms produced in the midgut penetrate the basement membrane, enter the body cavity, and infect other body tissues. In some pest species, viable spores in the killed midgut cells slough into the gut and are passed from the body in the fecal material. These spores remain viable and may be consumed by other immature insects feeding in or on the plant. The transfer of spores from one individual to an immature of the same species in the same generation is referred to as horizontal transmission.

Some of the microsporidium spores that enter the body cavity may infect reproductive tissues of both male and female hosts. Infections in the female develop in the ovaries and nearby cells and enter the ovum during maturation. The spores from such an infection pass from the female to the next generation via the egg. This type of disease transfer is called vertical transmission.

Hosts and Degree of Natural Control. Many subspecies of *B. thuringiensis* possess subtle differences in their crystalline proteins, enabling the subspecies to be lethal to specific insects. More than 45 subspecies of *Bt* are recognized universally. Some of these subspecies of *Bt* kill immature caterpillars and larvae in the order Diptera, and a few

Table 3. Bacteria and microsporidia attacking corn insects

Host insect	Pathogen	Occurrence[1]
Angoumois grain moth	*Bacillus thuringiensis*	A
Armyworms		
Armyworm	*Vairimorpha necatrix*	N
Fall armyworm	*Bacillus thuringiensis*	A
	Vairimorpha necatrix	A
Corn earworm	*Bacillus thuringiensis*	A
	Nosema heliothides	N
	Nosema sphingidis	A
	Bacillus thuringiensis	A
Southern corn rootworm	*Bacillus thuringiensis*	A
	Gregarina coronata	N
Cutworms		
Black cutworm	*Vairimorpha necatrix*	A
Variegated cutworm	*Bacillus thuringiensis*	N
European corn borer	*Nosema pyrausta*	N
	Vairimorpha necatrix	A
	Bacillus thuringiensis	N/A
	Bacillus megaterium	N
Fire ant (red imported)	*Thelohania* sp.	N
	Vairimorpha invictae	N
Grasshoppers		
Twostriped grasshopper	*Nosema acridophagus*	N
	Nosema cuneatum	N
	Nosema locustae	N
Differential grasshopper	*Nosema acridophagus*	N
	Nosema cuneatum	N
Fall webworm	*Nosema* sp.	A
	Vairimorpha necatrix	N/A
Japanese beetle	*Ovavesicular popilliae*	N
	Bacillus popilliae	N
	Bacillus thuringiensis	A
Mormon cricket	*Vairimorpha* sp.	A
Potato stem borer	*Nosema hydraeciae*	N
Sawtoothed grain beetle	*Nosema oryzaephili*	N
	Nosema whitei	N
Southwestern corn borer	*Bacillus thuringiensis*	A
Confused flour beetle	*Nosema whitei*	N
Red flour beetle	*Nosema oryzaephili*	N

[1] A, Applied as augmentative control; N, Natural control.

kill beetles and larvae in the order Hymenoptera. Natural control of insects by the soilborne Bacteriaceae seldom occurs.

Species in the genera *Nosema*, *Ovavesicular*, *Thelohania*, and *Vairimorpha* are single-celled eukaryotic organisms in the phylum Microspora and are present in many species of insects (Table 3). An infection by a microsporidium often causes an adult female insect to lay fewer eggs, and fewer eggs hatch. Resulting offspring are less vigorous, and, if undesirable environmental conditions exist, the growth of infected immatures is delayed or the infected insect dies.

Sampling Methods. Sampling for insects infected with these pathogens and detection of infected insects in the field are difficult.

Biological Control Potential. Commercial products containing *B. thuringiensis* are available for application to control several caterpillar pests of corn (see Table 3, Applied). Formulations of *Bt* sometimes are as effective as chemical insecticides for control of European corn borers. The microsporidia generally are not produced commercially, but they are passed efficiently within and between generations, often suppressing densities of pest insects.

Selected References. 31, 38, 92, 191, 241

Leslie C. Lewis

Nematodes

Scientific Classification. *Steinernema* (=*Neoaplectana*) and *Heterorhabditis* spp., Rhabditida (Steinernematidae and Heterorhabditidae).

Origin and Distribution. Eighteen *Steinernema* and eight *Heterorhabditis* species have been isolated from soil or invertebrate hosts found in North and South America, Europe, Australia, and Asia. Species such as *Steinernema carpocapsae* and *Heterorhabditis bacteriophora* have worldwide distributions. Four species from North and South America have been isolated from the corn earworm (*H. bacteriophora* and *S. riobravis*), fall armyworm (*S. riobravis*), Japanese beetle (*S. glaseri*), and a *Heliothis* species (*S. rara*).

Description and Biology. Entomopathogenic nematodes are microscopic, wormlike organisms that parasitize invertebrates. Free-living, infective-stage nematodes that are less than 1/20 in (1.2 mm) long and less than 1/50 in (0.5 mm) wide enter the insect host through a natural opening such as the mouth, trachea, or anus. Infective-stage nematodes invade the insect's body cavity and release bacteria (*Xenorhabdus* or *Photorhabdus* spp.) that are carried in the intestine of the nematode. The bacteria kill the insect in less than 48 hours and provide essential products for the development and reproduction of nematodes. The nematode's egg, four juvenile stages, and adult stages occur within the host insect. Reproduction can generate thousands of infective-stage nematodes that exit the host within 2 weeks after the infection. Parasitized insects typically turn light brown (*Steinernema* spp.) or red-brown (*Heterorhabditis* spp.) after death.

Hosts and Degree of Natural Control. Entomopathogenic nematodes are not pathogenic for mammals and birds but have an extensive host range for insects. Physical and environmental factors, such as insect size and sunlight, limit nematodes' ability to cause disease, especially in the natural habitat of the insect. Soil-associated stages of several corn insects (black cutworm, corn earworm, corn rootworms, European corn borer, fall armyworm, Japanese beetle) are susceptible to entomopathogenic nematodes. The ability to cause disease in insects varies among nematode species and geographic ranges.

Biological Control Potential. Entomopathogenic nematodes can be applied as aqueous suspensions of the infective stage with a ground sprayer or through overhead irrigation. Nematodes applied at 0.5–5 billion nematodes per acre (1.3–12 billion nematodes per hectare) have suppressed larval populations of corn rootworms and black cutworms. A natural population of *S. riobravis* has caused significant mortality to prepupae and pupae of corn earworms and fall armyworms in the Lower Rio Grande Valley. Improvements in application methods, selection of highly pathogenic species or isolates, and reduced costs of mass production of nematodes should enhance the potential for using entomopathogenic nematodes for control of several soil-inhabiting corn insects.

Selected References. 95, 136, 261

Jan J. Jackson

Parasitoids

Tachinid Flies

Scientific Classification. Several species, Diptera (Tachinidae).

Description and Life History. Females of the endoparasitic Tachinidae (those that live inside the host) deposit eggs or larvae directly into the host, eggs externally on the host, fully developed first instars on or near the host, or eggs on the food source of the host. Some tachinid larvae complete development and emerge from the host larvae to pupate; the development of others is accelerated and completed only after the host insect pupates. Adults are 1/5–1/2 in (4 –12 mm) long, depending upon species and the number of parasitoids produced per host. The Tachinidae typically parasitize mid- to late instars, thus contributing to the common perception that they have minor importance as biological control agents. Crop injury caused by the attacked host species may be marginally reduced or unaffected after parasitism of the host.

Tachinid fly eggs on caterpillar (Marlin E. Rice).

Hosts and Degree of Natural Control. The Tachinidae are the most important group of insect-feeding flies and generally are well represented among parasitoids collected from primary caterpillar pests of corn, particularly the European corn borer, corn earworm, fall armyworm, sugarcane borer, and black cutworm (Table 4).

Table 4. Primary tachinid parasitoids of lepidopteran pests of corn

Host	Tachinid parasitoid
Armyworms	
Armyworm	*Athrycia cinerea*
Fall armyworm	*Archytas marmoratus*
	Archytas incertus
	Eucelatoria rubentis
	Lespesia archippivora
	Lespesia aletia
	Winthemia trinitatis
	Winthemia rufopicta
Corn earworm	*Archytas marmoratus*
	Archytas incertus
	Eucelatoria rubentis
	Lespesia archippivora
	Lespesia aletia
Black cutworm	*Bonnetia comta*
	Lespesia archippivora
European corn borer	*Aplomya caesar*
	Lespesia archippivora
	Lixophaga diatraeae
	Lixophaga sp. near *variabilis*
	Lydella thompsoni
	Pseudoperichaeta nigrolineata
Sugarcane borer	*Lixophaga diatraeae*
Stalk borer	*Lixophaga thoracica*

Some tachinids attack a single species; others attack many species of pests. Their generally low rates of natural parasitism (from less than 1 to 30%) probably are attributable to unfavorable habitats, low numbers of parasitoids relative to the numbers of hosts, and the unavailability of hosts of appropriate size for parasitization. *Archytas marmoratus* in south Florida parasitized <1 and 17% medium and large fall armyworm larvae, respectively, in whorl-stage corn. This same species parasitized 37 and 58% medium and large larvae, respectively, in seven Atlantic and Gulf Coast states and in northeastern Mexico. The performance of this parasitoid can be improved only after the factors that influence these wide ranges in percentage parasitism are understood.

Biological Control Potential. Because tachinid flies are highly mobile, best results in percentage parasitism of pests have been obtained when flies were released in restricted areas, such as islands, or in fields surrounded by a physical barrier such as woodlands. In tests in south Georgia, inundative releases of *A. marmoratus* parasitized greater than 90% of the corn earworms in whorl-stage corn.

The development of economic and efficient methods for mass rearing high-quality parasitoids should enhance the use of tachinid parasitoids for managing caterpillar pests of corn. Rearing tachinid parasitoids on artificial diets has been somewhat successful. Inundative releases of tachinid parasitoids when seasonal generations of pests are beginning to develop are likely to be most effective in reducing pest densities.

Selected References. 29, 103, 239

Harry R. Gross

Braconid Wasps

Scientific Classification. Many species, Hymenoptera (Braconidae).

Origin and Distribution. Several species of braconids are important parasitoids of larvae of corn pests (Table 5). Although most are native to North America, the distribution of some species is not known. One species, *Macrocentrus grandii*, was introduced from Europe for control of European corn borers and is found in the midwestern and east-central states. Braconid wasps are not dangerous to humans.

Description and Life History. Braconids are small, usually dark wasps about 3/8 in (10 mm) or less in length, with long, curved antennae. Adult females search for larvae of their hosts into which they deposit one or more eggs. The eggs develop into small maggotlike larvae that feed

Aphid parasitized by braconid wasp (Marlin E. Rice).

inside the host larva. The host larva remains alive but grows slowly and may feed less than an nonparasitized larva. After the braconid larvae are full grown, they chew their way out of their host and spin silken cocoons. The presence of cocoons on or near the dead host larva is the most apparent evidence of attack by braconid wasps.

Although most braconids deposit eggs in the larval hosts, *Chelonus insularis* deposits a single egg in the egg stage of the fall armyworm. The wasp egg hatches and the wasp larva develops inside the fall armyworm larva.

Table 5. Primary braconid parasitoids of corn insect pests and reported parasitism rates

Host insect	Braconid parasitoid	% parasitism
Fall armyworm	*Chelonus insularis*[1]	<1–63 (US)[2]
		63 (FL)
		12 (SC)
	Aleiodes laphygamae	<1–2 (US)
		13 (GA)
	Cotesia marginiventris	<1–14 (US)
		6 (GA)
Corn earworm	*Microplitis croceipes*	1 (MS, TX)
Black cutworm	*Meteorus leviventris*	10–62 (IN)
European corn borer	*Macrocentrus grandii*	<1–45 (IL)
		1–34 (Eastern US)

FL, Florida; GA, Georgia; IL, Illinois; IN, Indiana; MS, Mississippi; SC, South Carolina; TX, Texas

[1] Parasitoid egg deposited in fall armyworm egg, parasitoid feeds in larva.

[2] When United States is reported, the range of percentage parasitism includes the results from more than one study site in the United States.

Hosts and Degree of Natural Control. Braconids are the most important of the many parasitoid species that attack the fall armyworm. Parasitized fall armyworms die as small- to medium-sized larvae. Although many fall armyworms are killed by braconid parasitoids, braconids do not maintain fall armyworm numbers below damaging levels in corn.

Macrocentrus grandii is a significant natural enemy of European corn borers in the midwestern and east-central United States. However, the decline in the level of parasitism by this species in some areas may have been caused by the increase of the microsporidian *Nosema pyrausta*, which causes a disease in both *M. grandii* and European corn borer.

Microplitis croceipes is a common parasitoid of corn earworms in many crops and regions. However, earworms feeding in corn ears apparently are inaccessible to the wasps and rarely are attacked by any parasitoid species.

Perilitis coccinellae is a common parasitoid of lady beetles that are important natural enemies of pests in corn.

Sampling Methods. The only way to determine if pest larvae are parasitized is to collect larvae of various ages and rear them to allow any parasitoids within to develop and emerge.

Effects of Management. An increase in the numbers and activity of parasitoids in agricultural crops could be facilitated by changing crop production practices; growing vegetation within fields and along field borders to provide alternative hosts, overwintering sites, and shelter; and providing food resources such as flowers for adult wasps. Also, avoiding the use of broad-spectrum insecticides applied to corn foliage may conserve braconids.

Biological Control Potential. Braconids are part of the complex of natural enemies that help suppress densities of several pests of corn. Development of farm management practices that enhance existing braconid populations and the introduction of exotic braconid species may increase the level of biological control provided by this group of parasitoids.

Selected References. 11, 146, 164, 193

Allen Knutson

Ichneumon Wasps

Scientific Classification. Many species, Hymenoptera (Ichneumonidae).

Origin and Distribution. The family of ichneumon wasps is one of the largest animal groups with an estimated 60,000 species worldwide and more than 3,100 described species in North America. Fifty-two species of Ichneumonidae attack pests of corn in North America, many of which are widely distributed. Ichneumon wasps are common in some parts of the world, comprising 5–8% of all insects species, and can be important regulators of insect populations. Ichneumon wasps are not dangerous to humans.

Ichneumon wasp (Marlin E. Rice).

Description and Life History. Adults are slender and vary from black to yellow-brown. They have long antennae with 16 or more segments. The female's ovipositor is permanently extended and usually is as long as or longer than the body. Adults vary from less than 1/8 in (3 mm) to 1 1/2 in (38 mm) long, not including the antennae and ovipositor.

Ichneumon wasps generally parasitize other insects that pass through egg, larva, pupa, and adult stages. Some species attack hosts in concealed locations such as stems or rolled leaves, immobilizing the hosts with venom. The wasp larvae then feed as external parasitoids. Other species insert an egg into the larva or pupa of the host. These wasp larvae feed on the internal body tissues of the host, eventually killing it.

Hosts and Degree of Natural Control. In corn, ichneumon wasps generally attack caterpillar pests such as the true armyworm (12 species of wasps; an example is *Campoletis flavicincta*), corn earworm (12 species of wasps; an example is *C. sonorensis*), European corn borer (17 species of wasps; an example is *Eriborus terebrans*), and cutworms (21 species of wasps; an example is *Hyposoter annulipes*). The imported ichneumonid *Lemophagus curtus* attacks cereal leaf beetle larvae. The level of natural parasitism by ichneumonids ranges widely, e.g., from 0 to nearly 60% for *E. terebrans* attacking corn borers. Some ichneumonids parasitize other beneficials; an example is *Isodromas lycaenae* attacking *Campoletis* and *Hyposoter* species.

Sampling Methods. Adult ichneumonids can be sampled in Malaise traps (large tentlike traps) and sticky traps or reared from parasitized hosts collected in the field. Some species can be collected around lights at night. Each of these methods has limitations for management decisions for biological control of corn pests.

Effects of Management. Hot and dry agricultural row-crop habitats, such as corn, are unfavorable for ichneumonid wasps. Because they are strong fliers, many ichneumonids leave unfavorable habitats. Ichneumonid wasps generally prefer cool, moist climates and habitats and require water and sugar from a combination of dew, plant nectar, and aphid honeydew.

Biological Control Potential. Conservation of ichneumonid wasps by managing agricultural habitats to provide food, water, shelter, and alternate hosts may increase the effectiveness of these parasitoids. Importation of additional species also might be effective.

Selected References. 43, 78, 104, 246

Douglas A. Landis

Trichogrammatid Wasps

Scientific Classification. Several *Trichogramma* spp., Hymenoptera (Trichogrammatidae); other egg parasitoids, Hymenoptera (Scelionidae and Mymaridae).

Origin and Distribution. Nine species of native *Trichogramma* wasps and a few other egg parasitoids are known from hosts on corn (Table 6). Many species probably remain to be identified, and the distributions of the known species are tentative. Two species, *T. ostriniae* from northern China and *T. brassicae* from Europe, are being introduced into North America. None of these wasps are dangerous to humans.

Description and Life History. Adult *Trichogramma* wasps are less than 1/32 in (0.7 mm) long, look like tiny ants, and are difficult to see. Adults primarily parasitize the eggs of Lepidoptera; immature stages develop inside the host eggs in 10–14 days. The presence of the parasitoid larva can be detected by the characteristic blackening of the "shell" of the host egg.

Hosts and Degree of Natural Control. *Trichogramma* wasps are opportunistic parasitoids, accepting most exposed, flat to spherical eggs, especially eggs of Lepidoptera. Potential hosts in corn include the European corn borer and corn earworm. Scelionids parasitize true bugs, and mymarids parasitize leafhoppers, although both parasitize some Lepidoptera. Some of these parasitoids attack eggs of other beneficial arthropods. The level of natural control varies from less than 1% to greater than 60% but usually is closer to the lower value (Table 6). None of these wasps maintain host populations below economic thresholds.

Table 6. Egg parasitoids of corn insects with their geographic distribution and reported parasitism rates

Host	Parasitoid	Distribution	% parasitism
	Trichogramma [Trichogrammatidae]		
Corn earworm	*T. minutum*	NA	<1 (TX)
	T. pretiosum	south of 42°N latitude	5–60 (TX, CA)
	T. exiguum		<10 (TX)
	T. maltbyi	TX, MI	<1 (TX)
	T. chilonis	HI	50 Jul–Oct, <20 Dec–June (HI)
European corn borer	*T. minutum*	NA	<1 (MN)
	T. nubilale	DE	5–70 (DE)
	T. exiguum	SC–MO & south	15 (SC)
Geocoris	*T. pretiosum*	south of 42°N latitude	<1 (AL)
Diatraea[1]	*T. pretiosum*	south of 42°N latitude	<5–33 (TX, KS, Mexico)
	T. deion	SD–TX & west and south	<5 (TX)
	T. atopovorilia	Mexico	60 (Jalisco, Mexico)
	Telenomus [Scelionidae]		
Corn earworm	*T. heliothidis*	TX, KS, LA	Trace–<10 (TX)
Geocoris	*T. reynoldsi*	AL, CA	10–60 (AL)
Diatraea[1]	*Telenomus* sp.	TX	<1 (TX)
	Mymaridae		
Corn earworm	mymarid sp.	TX	<1 (TX)

NA, Not Applicable; AL, Alabama; CA, California; DE, Delaware; HI, Hawaii; KS, Kansas; LA, Louisiana; MI, Michigan; MN, Minnesota; MO, Missouri; SC, South Carolina; SD, South Dakota; TX, Texas

[1] *Diatraea* is mainly *Diatraea grandiosella*, southwestern corn borer.

Sampling Methods. Because of their small size, the most efficient way of sampling trichogrammatid wasps is by collecting and rearing host eggs.

some species also feed on decaying plant material. Like ground beetles, rove beetles live in or near the soil surface, preferring areas with abundant decomposing plant residues. These beetles exhibit peak activity in spring and fall and at night.

Prey and Degree of Natural Control. Ground beetles feed on many pest and nonpest species. They prey primarily upon pests such as cutworms that spend a considerable portion of their life cycle in the soil. Studies in Iowa have indicated that ground beetles consumed as many as five cutworm larvae per day and as many as 13 cutworm eggs per day. Large densities of ground beetles have reduced the level of cutworm injury by as much as 98%. Because carabid beetles are active throughout much of the year and have generalized feeding habits, they help keep pest numbers below economic thresholds, rather than suppress pest outbreaks. In studies with the black cutworm in corn, ground beetles contributed more than 50% of predation-related mortality and a similar level of reduction in cutworm damage to corn.

Although little information about rove beetles in corn exists, these predators have been responsible for as much as a 30% reduction in densities of soil-inhabiting pests in other row crops.

Sampling Methods. Pitfall traps can be used to sample for ground and rove beetles. The beetles can be collected either alive or dead, depending on whether a preservative is included in the traps. Shallow soil samples, sticky boards, and systematic visual observations also have been used to sample for ground and rove beetles.

Effects of Management. Rearing ground or rove beetles for mass release is impractical, so provision of favorable habitat for these insects throughout the year is the best method of maximizing their benefits. Minimum tillage that leaves plant residues on or near the soil surface and planting cover crops help increase densities of ground beetles and other soil predators. Minimizing use of insecticides toxic to predatory beetles is another important component of management.

Selected References. 20, 35, 36, 236, 237

Benjamin R. Stinner and
David A. McCartney

Lady Beetles

Scientific Classification. Many species, Coleoptera (Coccinellidae).

Origin and Distribution. Three native species of lady beetles that occur in large numbers throughout North America are the twelvespotted lady beetle, *Coleomegilla maculata*; the convergent lady beetle, *Hippodamia convergens*; and the thirteenspotted lady beetle, *Hippodamia tredecimpunctata* (Table 8). The introduced sevenspotted lady beetle, *Coccinella septempunctata*, is a predominant species throughout the eastern United States. Several other species occur in low numbers, including *Adalia bipunctata*, *Cycloneda munda*, *C. polita*, *Hippodamia parenthesis*, *Coccinella novemnotata* (ninespotted lady beetle), *C. transversoguttata richardsoni* (transverse lady beetle), *Harmonia axyridis* (multicolored Asian lady beetle), and *Olla v-nigrum*. The genus *Stethorus* includes four species of mite-feeding lady beetles found primarily in the southern United States.

Description and Life History. The immature stages of lady beetles are not easily identified to species. Yellow to orange oval eggs (1–2 mm) typically are deposited in clusters; each egg is attached at the end to the underside of a corn leaf. The number of eggs per cluster varies from 2 to 70. Eggs hatch within 7–10 days; 12–24 hours before

Convergent lady beetle (left), twelvespotted lady beetle (center), sevenspotted lady beetle (right) (Marlin E. Rice).

hatching, eggs change from yellow to gray. Larvae often are black, brown, or dark blue, with orange or yellow stripes across the back. The larva's body, which often is spiny, is slender and tapered with well-developed legs. Larvae grow and molt through four instars; their developmental rate is dependent on temperature and the quality and quantity of food. Before pupation, the larva attaches its abdominal tip to a leaf or stem, curls its body into a C-shaped position, and pupates. Lady beetles have naked pupae (without cocoons). Pupae typically are orange to dark brown. Within a week, adults emerge from pupae. Initially their wing covers are soft, lightly yellow, and completely devoid of pattern or spots. The wing covers acquire their typical color

Table 8. Lady beetles associated wih corn, and their geographic distribution, description, and prevalence

Lady beetle species	Distribution	Body length in inches (mm)	Body shape, elytra color, number of spots	Prevalence in corn
Coleomegilla maculata	Ontario–GA, west to MN and TX	0.17–0.26 (4.2–6.6)	Broadly oval, pink to red, 12	Predominant
Hippodamia convergens	Entire United States, south Ontario, Manitoba, and British Columbia	0.17–0.29 (4.2–7.3)	Broadly oval, orange, variable (0–12)	Predominant
Hippodamia tredecim-punctata	Newfoundland to SC, west to AK and northern CA	0.18–0.25 (4.5–6.4)	Broadly oval, orange, 13	Predominant
Hippodamia parenthesis	Entire United States north of Gulf states	0.15–0.22 (3.8–5.6)	Broadly oval, orange, 5	Transient
Adalia bipunctata	Labrador to AL, west to AK and CA	0.14–0.21 (3.4–5.2)	Nearly round, red, 2	Transient
Coccinella septempunctata	Entire United States	0.26–0.31 (6.5–7.8)	Nearly round, orange to red, 7	Predominant in eastern United States
Cycloneda munda	East of Rockies	0.15–0.22 (3.7–5.7)	Nearly round, orange, 0	Transient
Cycloneda polita	Rockies to Pacific Coast	0.14–0.24 (3.5–6.2)	Nearly round, orange, 0	Transient
Coccinella novemnotata	ME to GA, west to British Columbia and CA	0.19–0.28 (4.7–7.0)	Nearly round, orange to red, variable (0–9)	Transient
Coccinella transversoguttata richardsoni	Labrador to VA, west to AL and CA	0.20–0.31 (5.0–7.8)	Nearly round, orange to red, 5 in bands	Transient
Harmonia axyridis	Eastern United States, WA	0.19–0.31 (4.8–7.5)	Nearly round, orange to red, variable (0–19)	Transient
Olla v-nigrum	Entire United States	0.15–0.24 (3.7–6.1)	Nearly round, pale yellow, 16	Transient
Stethorus spp.	Southern United States	<0.09 (<2.0)	Oval, black, 0	Transient

AL, Alabama; AK, Arkansas; CA, California; GA, Georgia; ME, Maine; MN, Minnesota; SC, South Carolina; TX, Texas; VA, Virginia; WA, Washington

Lady beetle larva (left), lady beetle eggs (right) (Marlin E. Rice).

patterns gradually, but the reddish color remains lighter for several days.

Photographs of some adult lady beetles have been provided to supplement their descriptions in Table 8. The scientific names of most species of lady beetles are based upon the distinguishing characteristics of the adults. For example, *H. convergens* is so named because it has two converging white lines on the pronotum.

Stethorus adults are not as colorful as the larger lady beetles. They are black with yellow antennae and mouthparts.

The seasonal life cycle of lady beetle species varies among geographic areas. In northern latitudes lady beetles may have 1–2 generations per season; 4 or more generations develop per season in southern latitudes.

Prey and Degree of Natural Control. Larvae and adults prey on aphids, other small insects, and insect eggs, particularly those of moths. In corn, they prey on corn leaf aphids; European corn borer, armyworm, and corn earworm eggs; and small soft-bodied insects, including mites and thrips. They also feed on corn pollen. The level of natural control is difficult to quantify, but lady beetles and other insect predators occasionally are responsible for the suppression of densities of corn leaf aphids and corn borer eggs to noneconomic levels.

Lady beetle pupa (Marlin E. Rice).

Sampling Methods. Yellow sticky traps and visual counts are used to sample for lady beetles. For better decision making in IPM in the future, the association of estimates of densities of lady beetles and the level of natural control they provide should be calculated.

Effects of Management. The use of resistant corn varieties does not seem to affect activities of lady beetles. Although intercropping probably favors an increase in the densities of many lady beetles, studies have indicated the opposite for *C. maculata*. Susceptibility of different species of lady beetles to insecticides varies greatly; larvae usually are more sensitive than pupae and adults.

Biological Control Potential. All three categories of biological control (importation, augmentation, and conservation) are applicable to lady beetles. *Coccinella septempunctata* is an imported lady beetle that now occurs in cornfields, preying on corn leaf aphids. Augmentation by releases of adult lady beetles is discouraged because beetles do not remain in fields. This dispersal behavior is observed in *H. convergens* adults that have been collected from overwintering sites, stored at low temperatures, and released. However, use of supplementary diets (e.g., dusting plants with pollen or sprays of sucrose solution, which has increased densities of lady beetle species in research plots) enhances naturally occurring species during periods of oviposition by European corn borers.

Selected References. 52, 101, 121, 262

M. W. Phoofolo, J. J. Obrycki, and G. J. Michels, Jr.

Flower Bugs and Minute Pirate Bugs

Scientific Classification. Several species in the genus *Orius*, Heteroptera (Anthocoridae).

Origin and Distribution. Two species are common in cornfields in North America. The insidious flower bug, *Orius insidiosus*, is more abundant in the eastern and central regions of the United States and Canada, whereas the minute pirate bug, *O. tristicolor*, is more common in the western regions of the United States and Canada. The two species have been introduced into Oahu, Hawaii. Other species, such as *Anthocoris whitei* and *Xylocoris discalis*, are extremely rare in cornfields. *Paratriphleps laevinsculus* was introduced from Peru to control the corn earworm in Texas, but the predator failed to establish.

Insidious flower bug nymph (left), insidious flower bug adult (right) (Marlin E. Rice)

Description and Life History. Adult *Orius* are 1/8 in (2–3 mm) long and appear black and white (black thorax and amber-clear wings). They overwinter as adults in leaf litter. *Orius* colonizes the whorls of foot-tall corn plants and later inhabits the leaf axils. Females embed their eggs in silks and leaf-collar tissues. The first four instars are oval and yellow-orange; the fifth (last) instar is dark brown. At 75°F (24°C), generation time from egg to adult is about 2 weeks. Peak *Orius* population density usually coincides with corn tasseling and pollen shedding.

Prey and Degree of Natural Control. Both nymphs and adults prey on thrips, mites, aphids, insect eggs, and young caterpillars (Table 9). Its mobility, small size, and large population densities enable *Orius* to prey effectively

Table 9. Flower bugs and minute pirate bugs associated with corn, reported predation rates, and relative abundance

Prey	Predator	Predation rate[1]	Peak density per plant[2]
Corn earworm (eggs)	Minute pirate bug[3]	<20 (CA)	0.1[4] (CA)
	Insidious flower bug[5]	up to 52 (NC)	3.2 (NC)
	Insidious flower bug	up to 22 (MD)	0.8–1.7 (MD)
	Insidious flower bug	14–54 (VA)	0.8 6.8 (VA)
	Insidious flower bug		4.6 (GA)
	Insidious flower bug		6.6 (GA)
	Insidious flower bug	25–50 (KS)	
Corn earworm (eggs and larvae)	Insidious flower bug		6.2–7.5 (MO)
Corn leaf aphid	Insidious flower bug		1.5–4.0 (MD)
	Insidious flower bug		2.4–13.0 (IL)
European corn borer (eggs)	Insidious flower bug	4–19 (MN)	
	Insidious flower bug	20 (MD)	1.5–4.0 (MD)
European corn borer (larvae)	Insidious flower bug	5 (MD)	1.5–4.0 (MD)
European corn borer (eggs and larvae)	Insidious flower bug		1.1 (NE)
	Insidious flower bug		2.6–8.2 (OH)
	Insidious flower bug		7.8–19.6 (OH)
	Insidious flower bug		0.1–2.4 (IA)
	Insidious flower bug		0.5–2.3 (MA)
Southwestern corn borer (eggs and larvae)	Insidious flower bug	<38 (TX)	4.8 (TX)
Spider mites (*Oligonychus pratensis, Tetranychus urticae*)	Insidious flower bug		0.5–4.5 (TX)
Thrips (*Frankliniella williamsi, F. tenuicornis*)	Insidious flower bug		1.5–4.0 (MD, OH)

CA, California; GA, Georgia; IL, Illinois; IA, Iowa; KS, Kansas; KY, Kentucky; MD, Maryland; MA, Massachusetts; MO, Missouri; NE, Nebraska; NC, North Carolina; OH, Ohio; TX, Texas; VA, Virginia

[1] Percentage of sampled prey species killed by the predator.

[2] Number of bugs per plant at peak population density (ranges indicate values in different years or fields).

[3] *Orius tristicolor*.

[4] D-vac sample.

[5] *Orius insidiosus*.

on many concealed corn pests, even when densities of prey are low. In addition, scarcity of prey may not decrease *Orius* density because the predator can sustain itself on corn pollen and silk. Feeding and oviposition by *Orius* in corn silk do not seem to have adverse effects on pollination.

Sampling Methods. Because *Orius* often is concealed in whorls, leaf axils, and ears, a reliable estimate of the predator's density requires destructive sampling of plants. Density of *Orius* may be estimated by examining fresh silks at pollen shedding. *Orius* often is abundant on corn varieties with several ears and profuse silk production and in weedy fields.

Effects of Management. Insecticide application often is detrimental to *Orius* populations. Because of its plant-feeding habits, *Orius* also may be affected adversely by systemic insecticides.

Biological Control Potential. In corn, *Orius* is most effective against the corn earworm (Table 9). Although several *Orius* species are available commercially, their use in corn is limited by high costs. Therefore, conservation of naturally occurring populations is the most feasible way to enhance biological control of corn pests by *Orius*. *Orius* activity could be encouraged by introducing pollen-producing plants in or around cornfields (intercropping, weedy field edges) and by minimizing the use of harmful pesticides.

Selected References. 117, 122, 131, 197

Moshe Coll

Damsel Bugs and Stink Bugs

Scientific Classification. Several species of damsel bugs, Heteroptera (Nabidae), and stink bugs, Heteroptera (Pentatomidae).

Origin and Distribution. Some common damsel bugs in North America, and their distribution in North America, are *Nabis alternatus* (western), *N. americoferus* (primarily northern), *N. capsiformis* (southern), and *N. roseipennis* (primarily eastern). Damsel bugs are plentiful in many crops, although they are not common in corn.

Stink bugs are found throughout North America, including the predators *Perillus bioculatus*, *Podisus maculiventris* (the spined soldier bug), and *Stiretrus anchorago*. Predaceous stink bugs are found in most crop habitats, but they are not abundant in corn.

Description and Life History. Adult damsel bugs are 1/5–1/2 in (5–12 mm) long and slender, with slightly enlarged front legs. Most common damsel bugs are either mottled brown, pale yellow, or black. Both long- and short-winged forms occur. Damsel bugs are predaceous, but they also feed on plants. Adult damsel bugs deposit eggs in leaves and plant stems. Nymphs are similar in shape to adults.

Predaceous stink bugs are broad, shield-shaped, and 1/3–3/4 in (7–20 mm) long. Most species are brown on top and paler below, although some are brightly colored. Eggs usually are barrel shaped, deposited in clusters, and glued to the plant surface. They generally hatch within a week. Predaceous and phytophagous stink bugs are difficult to distinguish; predaceous *Podisus* species resemble the phytophagous pest *Euschistus*. Predaceous species have thicker, stronger front legs and thicker, stouter mouthparts than phytophagous species.

Hosts and Degree of Natural Control. Damsel bugs and stink bugs are opportunistic, generalist predators. They attack many small prey species within the plant canopy. Damsel bugs attack eggs and larvae of several Lepidoptera (armyworm, corn earworm, and European corn borer), leafhoppers, thrips, aphids, and chinch bugs. Stink bugs attack caterpillars (corn earworm, armyworm, black

Spined soldier bugs (Marlin E. Rice).

Damsel bug adult (left) and nymph (right) (Marlin E. Rice).

cutworm) and other slow-moving prey. These predators likely have considerably less impact than *Orius* on pests in corn. In other crops, both damsel bugs and predatory stink bugs provide natural control of many pests.

Biological Control Potential. Damsel bugs warrant closer examination to determine their true potential.

Selected References. 5, 28, 64, 215

Robert N. Wiedenmann

Green and Brown Lacewings

Scientific Classification. Several species of green lacewings, Neuroptera (Chrysopidae), and brown lacewings, Neuroptera (Hemerobiidae).

Origin and Distribution. The Chrysopidae and Hemerobiidae are the second and third largest families of Neuroptera and include 87 and 58 North American species, respectively. Chrysopidae include about 1,200 species and subspecies in 86 genera and subgenera worldwide. Some green and brown lacewing species are common and widely distributed; others are relatively rare and have restricted distributions. *Chrysoperla carnea*, one of the most common species, is found throughout much of the temperate regions of the world, except in Australia. *Hemerobius pacificus* is restricted to western North America.

Description and Life History. Adult green lacewings are about 1 in (25 mm) long and generally are green or yellow, although a few species are dark colored. Adult brown lacewings are brown or yellow-brown and generally are smaller than green lacewings. Adults of both families have large, lacy wings. Green lacewings have threadlike antennae; brown lacewings have antennae that resemble strings of beads. Larvae are elongate and flat and have large sickle-shaped mandibles that they use to pierce prey and suck body fluids. Eggs of green lacewings generally are deposited on top of a slender silk stalk and are laid singly, in loose groups, or in clusters, depending on the species. Brown lacewing eggs are deposited directly on the plant. The adults of many green lacewings are not predaceous; rather, they feed on honeydew, flower nectar, and pollen. Both larvae and adult brown lacewings are predaceous. Green lacewings and brown lacewings are predominantly arboreal, although some species, including *C. carnea* and *C. rufilabris*, are widespread in occurrence and frequently are found in field crops such as corn. These species move about freely and prey on many species of insects. Although adults are weak flyers, they can be transported long distances by the wind.

Prey and Degree of Natural Control. Lacewing larvae are effective searchers that feed on many small, soft-bodied insects, spiders, and mites and on insect eggs. They are general predators and do not establish a density-dependent (regulatory) relationship with any prey species. In other words, the density of lacewings is not dependent upon the density of any prey species.

Green lacewing (left) and green lacewing egg (right) (Marlin E. Rice).

Brown lacewing (Marlin E. Rice).

Lacewings comprise a small part of the complex of predators in crops, especially early in the season. Green lacewings require warmer temperatures for development than brown lacewings. Thus, green lacewings seldom become abundant until late in the season. Also, lacewing larvae and pupae are attacked by many predators and parasitoids, resulting in significant mortality.

Sampling Methods. Adult lacewings are sampled with sweep nets and light traps to determine species composition or to observe seasonal changes in population density. Larvae are sampled by beating vegetation over a tray or drop cloth, or with a sweep net. Green lacewing eggs perched on the top of stalks are easy to recognize. Egg counts on a random sample of plants provide estimates of population density.

Green lacewing larva (Marlin E. Rice).

Green lacewing larva (Marlin E. Rice).

Effects of Management. Compared with other predators, some important green lacewings are relatively tolerant to many conventional pesticides, including pyrethroids. Irrigation may enable lacewings to persist during dry periods. Mowing and its disruption of the microclimate may affect lacewing populations adversely. Habitats that support lacewing populations (reservoir habitats) include forested and weedy areas, where lacewings diapause, and other crops. The distance of crops from these reservoirs influences the extent of invasion by lacewings into the crops. Thus, diverse agroecosystems, with some areas of trees or weeds, or systems with a continuous sequence of crops may be beneficial for these predators.

Biological Control Potential. Although green lacewings are common, efficient, and often important predators, they seldom provide adequate natural control, particularly in field crops. However, their populations in agricultural systems can be increased by periodic releases and environmental manipulation. Periodic releases of green lacewings such as *C. carnea* and *C. rufilabris*, which are available commercially, have provided effective control of pests in some crops, including vegetables, fruits, potatoes, and cotton. However, little data are available on their effectiveness in corn. Some studies have demonstrated that sucrose sprays in corn increase the level of predation of European corn borers by *C. carnea*.

Selected References. 2, 32, 41, 44

Donald A. Nordlund

References Cited

1 **Adams, R. G., and L. M. Los. 1986.** Monitoring adult corn flea beetles (Coleoptera: Chrysomelidae) in sweet corn fields with color sticky traps. Environ. Entomol. 15: 867-873.

2 **Agnew, C. W., W. L. Sterling, and D. A. Dean. 1981.** Notes on the Chrysopidae and Hemerobiidae of eastern Texas with keys for their identification. Southwestern Entomol., Suppl. No. 4.

3 **Ainslie, G. G. 1922.** Webworms injurious to cereal and forage crops and their control. USDA Farmers Bull. 1258: 3–16, Washington, D. C.

4 **Alexander, C. P., and G. W. Byers. 1981.** Tipulidae, pp. 153-190. *In* J. F. McAlpine, B.V. Peterson, G. E. Shewell, H. J. Teskey, J. R. Vockeroth, and D. M. Wood [eds.], Manual of Nearctic Diptera, vol. 1. Agriculture Canada, Hull, Quebec.

5 **Alomar, O., and R. N. Wiedenmann [eds.]. 1996.** Zoophytophagous Heteroptera: implications for life history and integrated pest management. Thomas Say Publications in Entomology, Entomol. Soc. Am., Lanham, Maryland.

6 **Altieri, M. A., and M. Liebman [eds.]. 1988.** Weed management in agroecosystems: ecological approaches. CRC Press, Boca Raton, Florida.

7 **Altieri, M. A., and W. H. Whitcomb. 1980.** Weed manipulation for insect pest management in corn. Environ. Management 4: 483–489.

8 **Amrine, J. W., Jr., and T. A. Stasny. 1994.** Catalog of the Eriophyoidea (Acarina: Prostigmata) of the world. Indira Publishing Co., West Bloomfield, Michigan.

9 **Annan, I. B., and M. K. Bergman. 1988.** Effects of the onespotted stink bug (Hemiptera: Pentatomidae) on growth and development of corn. J. Econ. Entomol. 81: 649–653.

10 **Appel, L. L., R. J. Wright, and J. B. Campbell. 1993.** Economic injury levels for western bean cutworm, *Loxagrotis albicosta* (Smith) (Lepidoptera: Noctuidae), eggs and larvae in field corn. J. Kansas Entomol. Soc. 66: 434–438.

11 **Ashley, T. R. 1986.** Geographic distributions and parasitization levels for parasitoids of the fall armyworm, *Spodoptera frugiperda*. Florida Entomol. 69: 516–525.

12 **Ashlock, P. D. 1977.** New records and name changes of North American Lygaeidae (Hemiptera: Heteroptera: Lygaeidae). Proc. Entomol. Soc. Wash. 79: 575–582.

13 **Ashton, F. M., and T. J. Monaco. 1991.** Weed science: principles and practices, 3rd ed. John Wiley & Sons, New York.

14 **Ayre, G. L., and R. J. Lamb. 1990.** Life histories, flight patterns, and relative abundance of nine cutworms (Lepidoptera: Noctuidae) in Manitoba. Can. Entomol. 122: 1059–1070.

15 **Bailey, W. C., and L. Pedigo. 1986.** Damage and yield loss induced by stalk borer (Lepidoptera: Noctuidae) in field corn. J. Econ. Entomol. 79: 233–237.

16 **Baker, J. R. 1982.** Insects and other pests associated with turf: some important, common, and potential pests in the southeastern United States. North Carolina Agricultural Extension Service AG-268, Raleigh.

17 **Barber, H. G. 1947.** Revision of the genus *Nysius* in the United States and Canada (Hemiptera: Heteroptera: Lygaeidae). J. Wash. Acad. Sci. 37: 354–366.

18 **Berry, R. E. 1978.** Insects and mites of economic importance in the Northwest. Oregon State University Book Stores, Corvallis.

19 **Beshear, R. J. 1979.** Additional records of thrips in Georgia (Thysanoptera: Terebrantia). J. Georgia Entomol. Soc. 14: 209–211.

20 **Best, R. L., and C. C. Beegle. 1977.** Consumption of *Agrotis ipsilon* by several species of carabids found in Iowa. Environ. Entomol. 6: 532–534.

21 **Bing, L. A., and L. C. Lewis. 1992.** Endophytic *Beauveria bassiana* (Balsamo) Vuillemin in corn: the influence of the plant growth stage and *Ostrinia nubilalis* (Hübner). Biocontrol 2: 39–47.

22 **Bing, J. W., W. D. Guthrie, F. F. Dicke, and J. J. Obrycki. 1991.** Seedling stage feeding by corn leaf aphid (Homoptera: Aphididae): influence on plant development in maize. J. Econ. Entomol. 84: 625–632.

23 **Blackman, R. L., and V. F. Eastop. 1984.** Aphids on the world's crops. John Wiley & Sons, New York.

24 **Blackmer, J. L., and G. W. Bishop. 1991.** Population dynamics of *Rhopalosiphum padi* in corn in relation to barley yellow dwarf epidemiology in southwestern Idaho. Environ. Entomol. 20: 166–174.

25 **Blake, D. H. 1942.** Beetles of the genus *Luperodes*. Proc. U. S. National Museum 92: 67–68.

26 **Boote, K. J. 1981.** Concepts for modeling crop response to pest damage. Am. Soc. Agric. Eng. Paper 81: 4007.

27 **Brakke, M. K. 1987.** Wheat streak mosaic virus, pp. 589–590. *In* Wheat and wheat improvement. Agronomy Monograph No. 13, 2nd ed. American Society of Agronomy, Crop Science Society of America, and Soil Science Society of America, Madison, Wisconsin.

28 **Braman, S. K., and K. V. Yeargan. 1988.** Comparison of developmental and reproductive rates of *Nabis americoferus*, *N. roseipennis*, and *N. rufusculus* (Hemiptera: Nabidae). Ann. Entomol. Soc. Am. 81: 923–930.

29 **Bratti, A., and W. C. Nettles, Jr. 1992.** *In vitro* rearing of *Eucelatoria bryani*: improvements and evaluation of factors affecting efficiency. Entomol. Exp. Appl. 63: 213–219.

30 **Bridges, D. C. [ed.]. 1992.** Crop losses due to weeds in the United States. Weed Sci. Soc. Am., Champaign, Illinois.

31 **Brooks, W. M. 1985.** Entomogenous Protozoa, pp. 1 150. *In* C. M. Ignoffo [ed.], CRC handbook of natural pesticides, vol. V - microbial insecticides, part A - entomogenous protozoa and fungi. CRC Press, Boca Raton, Florida.

32 **Brooks, S. J., and P. C. Barnard. 1990.** The green lacewings of the world: a generic review. Bull. British Museum Nat. Hist. (Entomol.) 59: 117–286.

33 **Brooks, H. L., and R. A. Higgins. 1995.** Sorghum insect management for 1995. MF-742 (Revised). Cooperative Extension Service, Kansas State University, Manhattan.

34 **Browning, H. W., M. O. Way, and B. M. Drees. 1989.** Managing the Mexican rice borer in Texas. Texas Agricultural Experiment Station Bulletin B-1620, College Station.

35 **Brust, G. E., B. R. Stinner, and D. A. McCartney. 1985.** Tillage and soil insecticide effects on predator-black cutworm (Lepidoptera: Noctuidae) interactions in corn agroecosystems. J. Econ. Entomol. 78: 1389–1392.

36 **Brust, G. E., B. R. Stinner, and D. A. McCartney. 1986.** Predator activity and predation in corn agroecosystems. Environ. Entomol. 15: 1017–1021.

37 **Buntin, G. D., L. P. Pedigo, and W. B. Showers. 1990.** Temporal occurrence of the variegated cutworm (Lepidoptera: Noctuidae) adults in Iowa with evidence of migration. Environ. Entomol. 19: 603–608.

38 **Burges, H. D. 1981.** Microbial control of pests and plant diseases 1970–1980. Academic, New York.

39 **Byers, R. A. 1965.** Biology and control of a spittlebug, *Prosapia bicincta* (Say), on coastal bermudagrass. Georgia Agricultural Experiment Station Technical Bulletin n.s. 42, Athens.

40 **Caceres, O. 1992.** Age determination of preserved European corn borer larvae (Lepidoptera: Pyralidae) with notes on shank versus stalk tunneling of field corn. M.S. thesis, Kansas State University Library, Manhattan.

41 **Canard, M., Y. Séméria, and T. R. New. 1984.** Biology of Chrysomelidae. Vol. 27, Series Entomologica. W. Junk, The Hague, Netherlands.

42 **Capinera, J. L., and T. S. Sechrist. 1982.** Grasshoppers (Acrididae) of Colorado. Colorado State University Experiment Station Bulletin 584S, Ft. Collins.

43 **Carlson, R. W. 1979.** Family Ichneumonidae, pp. 315–740. *In* K. V. Krombein et al. [eds.], Catalog of Hymenoptera in America north of Mexico, vol. 1. Smithsonian Institution Press, Washington, D. C.

44 **Carlson, R. E., and H. C. Chiang. 1973.** Reduction of an *Ostrinia nubilalis* population by predatory insects attracted by sucrose sprays. Entomophaga 18: 205–211.

45 **Cave, R. D., and M. J. Gaylor. 1988.** Parasitism of *Geocoris* (Hemiptera: Lygaeidae) eggs by *Telenomus reynoldsi* (Hymenoptera: Scelionidae) and *Trichogramma pretiosum* (Hymenoptera: Trichogrammatidae) in Alabama. Environ. Entomol. 17: 945–951.

46 **Chiang, H. C. 1973.** Bionomics of the northern and western corn rootworms. Annu. Rev. Entomol. 18: 47–72.

47 **Chippendale, G. M. 1979.** The southwestern corn borer, *Diatraea grandiosella*: case history of an invading insect. Missouri Agricultural Experiment Station Research Bulletin 1031, Columbia.

48 **Chittenden, F. H. 1901.** The fall armyworm and the variegated cutworm. USDA Bull. No. 29, new series, Washington, D. C.

49 **Chittenden, F. H. 1916.** The pink corn-worm: an insect destructive to corn in the crib. USDA Bull. 363, Washington, D. C.

50 **Clausen, C. P. [ed]. 1978.** Introduced parasites and predators of arthropod pests and weeds: a world review. Agricultural Handbook No. 480, Agricultural Research Service, USDA, Washington, D. C.

51 **Coceano, P. G., and S. Peressini. 1989.** Colonization of maize by aphid vectors of barley yellow dwarf virus. Ann. Appl. Biol. 114: 443–447.

52 **Coderre, D., L. Provencher, and J. C. Tourneur. 1987.** Oviposition and niche partitioning in aphidophagous insects on maize. Can. Entomol. 119: 195–203.

53 **Cook, W. C. 1930.** Field studies of the pale western cutworm (*Porosagrotis orthogonia* Morr.). University of Montana Bulletin 225, Billings.

54 **Cowan, F. T. 1929.** Life history, habits and control of the Mormon cricket. USDA Tech. Bull. 161, Washington, D. C.

55 **Crumb, S. E. 1926.** The bronzed cutworm (*Nephelodes emmedonia* Cramer) (Lepidoptera). Entomol. Soc. Wash. Proc. 28: 201–207.

56 **Crumb, S. E. 1929.** Tobacco cutworms. USDA Tech. Bull. 88, Washington, D. C.

57 **Davidson, R. H., and W. F. Lyon. 1987.** Insect pests of farm, garden, and orchard. John Wiley & Sons, New York.

58 **Davis, J. J. 1917.** The corn root aphid and methods of controlling it. USDA Farmers' Bull. 891, Washington, D. C.

59 **Davis, P. M., and L. P. Pedigo. 1990.** Evaluation of two management strategies for stalk borer, *Papaipema nebris*, in corn. Crop Protection 9: 387–391.

60 **Davis, P. M., and L. P. Pedigo. 1991.** Economic injury levels for management of stalk borer (Lepidoptera: Noctuidae) in corn. J. Econ. Entomol. 84: 290–293.

61 **Decker, G. C. 1931.** The biology of the stalk borer *Papaipema nebris* (Gn.). Iowa Agricultural Experiment Station Research Bulletin 143: 289–351, Ames.

62 **Decker, G. C., and J. V. Maddox. 1971.** Effect of temperature on rate of development and survival of *Simyra henrici*. J. Econ. Entomol. 64: 94–98.

63 **Decker, G. C., and J. V. Maddox. 1971.** Observations on the bionomics of *Simyra henrici*. J. Econ. Entomol. 64: 117–122.

64 **DeClercq, P., and D. Degheele. 1990.** Description and life history of the predatory bug *Podisus saggita* (Fab.) (Hemiptera: Pentatomidae). Can. Entomol. 122: 1149–1156.

65 **Deedat, Y. D., and C. R. Ellis. 1983.** Damage caused by potato stem borer (Lepidoptera: Noctuidae) to field corn. J. Econ. Entomol. 76: 1055–1060.

66 **Deedat, Y. D., C. R. Ellis, and R. J. West. 1983.** Life history of the potato stem borer (Lepidoptera: Noctuidae) in Ontario. J. Econ. Entomol. 76: 1033–1037.

67 **Dethier, V. G. 1987.** The feeding behavior of a polyphagous caterpillar (*Diacrisia virginica*) in its natural habitat. Can. J. Zool. 66: 1280–1288.

68 **Dicke, F. F., and W. D. Guthrie. 1988.** The most important corn insects, pp. 767-867. *In* G. F. Sprague and J. W. Dudley [eds.], Corn and corn improvement. Agronomy Monograph No. 18, 3rd ed. American Society of Agronomy, Crop Science Society of America, and Soil Science Society of America, Madison, Wisconsin.

69 **Dill, J. F. 1979.** Biology and management of the corn flea beetle, *Chaetocnema pulicaria* Melsheimer, relative to the incidence of Stewart's disease in corn. Ph.D. Dissertation. Purdue University, W. Lafayette, Indiana.

70 **Dolbeer, R. A. 1980.** Blackbirds and corn in Ohio. U. S. Fish & Wildlife Service Research Publication 38, Washington, D. C.

71 **Dolbeer, R. A. 1990.** Ornithology and integrated pest management: red-winged blackbirds, *Agelaius phoeniceus*, and corn. Ibis 132: 309–322.

72 **Dolbeer, R. A., N. R. Holler, and D. W. Hawthorne. 1994.** Identification and control of wildlife damage, pp. 474–506. *In* T. A. Bookhout [ed.], Research and management techniques for wildlife and habitats. The Wildlife Society, Bethesda, Maryland.

73 **Dolbeer, R. A., P. P. Woronecki, and T. W. Seamans. 1995.** Ranking and evaluation of field corn hybrids for resistance to blackbird damage. Crop Protection 14: 399–403.

74 **Drees, B. M. 1987.** Control of southern corn rootworm with at-plant soil insecticides. Down to Earth 43: 7–11.

75 **Drees, B. M., and S. B. Vinson. 1992.** Fire ants and their management. B-1536, Texas Agricultural Extension Service, Texas A & M University System, College Station.

76 **Drees, B. M., L. A. Berger, R. Cavazos, and S. B. Vinson. 1991.** Factors affecting sorghum and corn seed predation by foraging red imported fire ants (Hymenoptera: Formicidae). J. Econ. Entomol. 84: 285–289.

77 **Dwinardi, A., L. H. Townsend, and J. D. Sedlacek. 1989.** Yield reduction from feeding by *Euschistus servus* and *E. variolarius* (Heteroptera: Pentatomidae) on stage V2 corn. J. Econ. Entomol. 82: 445–448.

78 **Dyer, L. E., and D. A. Landis. 1996.** Effects of habitat, temperature and sugar availability on longevity of *Eriborus terebrans* (Gravenhorst) (Hymenoptera: Ichneumonidae). Environ. Entomol. 25: 1192–1201.

79 **Eastman, C. E. 1980.** Sampling phytophagous underground soybean arthropods, pp. 327–354. *In* M. Kogan and D. C. Herzog [eds.], Sampling methods in soybean entomology. Springer-Verlag, New York.

References Cited

80 **Eddy, C. O. 1940.** Entomological progress, No. 2. Bulletin of the Louisiana Agricultural Experiment Station No. 323, Baton Rouge.

81 **Edwards, C. R., J. L. Obermeyer, T. N. Jordan, D. J. Childs, D. H. Scott, J. M. Ferris, R. M. Corrigan, and L. W. Bledsoe. 1992.** Japanese beetle, p. B-109. *In* Field crops pest management manual, IPM-1. Purdue University Cooperative Extension Service, W. Lafayette, Indiana.

82 **Edwards, C. R., J. L. Obermeyer, T. N. Jordan, D. J. Childs, D. H. Scott, J. M. Ferris, R. M. Corrigan, M. K. Bergman, R. H. Johnson, and A. C. Sayers. 1992.** Seed corn pest management manual for the Midwest, IPM-2. Purdue University Cooperative Extension Service, W. Lafayette, Indiana.

83 **Elliott, N. C., and R. W. Kieckhefer. 1989.** Effects of constant and fluctuating temperatures on immature development and age-specific life tables of *Rhopalosiphum padi* (L.). Can. Entomol. 121: 131–140.

84 **Esselbaugh, C. O. 1948.** Notes on the bionomics of some Midwestern Pentatomidae. Entomologica Americana 28: 1–73.

85 **Essig, E. O. 1934.** Dermaptera, pp. 63–65. *In* Insects of western North America. The MacMillan Co., New York.

86 **Eyer, J. R. 1954.** The pale western cutworm in eastern New Mexico. New Mexico Experiment Station Bulletin 414, Las Cruces.

87 **Fleming, W. E. 1970.** The Japanese beetle in the United States. USDA, Agricultural Research Service, Agriculture Handbook No. 236, Washington, D. C.

88 **Flint, M. L. 1990.** Pests of the garden and small farm. Statewide IPM Project, University of California, Division of Agriculture and Natural Resources, Publication 3332, Davis.

89 **Flynn, J. L., T. E. Reagan, and E. O. Ogunwolu. 1984.** Establishment and damage of the sugarcane borer (Lepidoptera: Pyralidae) in corn as influenced by plant development. J. Econ. Entomol. 77: 691–697.

90 **Foott, W. H., and P. R. Timmins. 1973.** Effects of infestations by the corn leaf aphid, *Rhopalosiphum maidis* (Homoptera: Aphididae), on field corn in Ontario. Can. Entomol. 105: 449–458.

91 **Forbes, S. A. 1904.** The more important insect injuries to Indian corn. Illinois Agricultural Experiment Station Bulletin 95: 331–339, Urbana-Champaign.

92 **Fuxa, J. R., and Y. Tanada.** 1987. Epizootiology of insect diseases. John Wiley & Sons, New York.

93 **Gardner, W. A., and J. R. Fuxa. 1980.** Pathogens for the suppression of the fall armyworm. Florida Entomol. 63: 439–446.

94 **Gardner, W. A., R. Noblet, and R. D. Schwehr. 1984.** The potential of microbial agents in managing populations of the fall armyworm (Lepidoptera: Noctuidae). Florida Entomol. 67: 325–332.

95 **Gaugler, R., and H. K. Kaya [eds.]. 1990.** Entomopathogenic nematodes in biological control. CRC Press, Boca Raton, Florida.

96 **Gerber, G. H., and J. Walkof. 1992.** Phenology and reproductive status of adult redbacked cutworms, *Euxoa ochrogaster* (Guenee) (Lepidoptera: Noctuidae), in southern Manitoba. Can. Entomol. 124: 541–551.

97 **Gibb, T., and R. Higgins. 1991.** Aboveground dry weight and yield response of field corn to defoliation and root pruning stresses. J. Econ. Entomol. 84: 1562–1576.

98 **Giebink, B. L., J. M. Scriber, and J. L. Wedberg. 1984.** Biology and phenology of the hop vine borer, *Hydraecia immanis* Guenee, and detection of the potato stem borer, *H. micacea* (Esper) (Lepidoptera: Noctuidae), in Wisconsin. Environ. Entomol. 13: 1216–1224.

99 **Giebink, B. L., J. M. Scriber, and D. B. Hogg. 1985.** Developmental rates of the hop vine borer and potato stem borer (Lepidoptera: Noctuidae): implications for insecticidal control. J. Econ. Entomol. 78: 311–315.

100 **Godan, D. 1983.** Pest slugs and snails, biology and control. Springer-Verlag, New York.

101 **Gordon, R. D. 1985.** The Coccinellidae (Coleoptera) of America north of Mexico. J. New York Entomol. Soc. 93: 1-912.

102 **Gordon, D. T., J. K. Knoke, and G. E. Scott [eds.]. 1981.** Virus and viruslike diseases of maize in the United States. Southern Cooperative Series Bulletin 247, Ohio State University, Columbus.

103 **Gross, H. R., Jr. 1990.** Field release and evaluation of *Archytas marmoratus* (Diptera: Tachinidae) against larvae of *Heliothis zea* (Lepidoptera: Noctuidae) in whorl stage corn. Environ. Entomol. 19: 1122–1129.

104 **Gupta, V., and S. Gupta. 1991.** Bibliography of Ichneumonidae. Memoirs Am. Entomol. Inst. 48: 1–355.

105 **Haack, K. D., and T. A. Granovsky. 1990.** Ants, pp. 415–479. *In* A. Mallis [ed.], Handbook of pest control. 7th ed. Franzak & Foster Co., Cleveland, Ohio.

106 **Hagen, A. F. 1979.** Western bean cutworm in corn. NebGuide G 76-290. Cooperative Extension Service, University of Nebraska, Lincoln.

107 **Hammond, R. B. 1996.** Conservation tillage and slugs in the U. S. corn belt, paper no. 66. *In* I. F. Henderson, Chair, Proc. BCPC Symposium: Slug and snail pests in agriculture. British Crop Protection Council., Farnham, Surrey, England.

108 **Hammond, R. B., and R. L. Cooper. 1993.** Interaction of planting times following the incorporation of a living, green cover crop and control measures on seedcorn maggot populations in soybean. Crop Protection 12: 539–543.

109 **Hammond, R. B., and B. R. Stinner. 1987.** Seedcorn maggots (Diptera: Anthomyiidae) and slugs in conservation tillage systems in Ohio. J. Econ. Entomol. 80: 680–684.

110 **Hammond, R. B., J. A. Smith, and T. Beck. 1996.** Timing of molluscicide applications for reliable control in no-tillage field crops. J. Econ. Entomol. 89:1028–1032.

111 **Hardwick, D. F. 1965.** The ochrogaster group of the genus *Euxoa* (Lepidoptera: Noctuidae), with description of a new species. Can. Entomol. 97: 673–678.

112 **Harned, R. W. 1915.** The corn-silk beetle, *Luperodes varicornis* LeC., and its control. J. Econ. Entomol. 8: 507–508.

113 **Harvey, T. L., K. D. Kofoid, T. J. Martin, and P. E. Sloderbeck. 1991.** A new greenbug virulent to E-biotype resistant sorghum. Crop Sci. 31: 1689–1691.

114 **Harvey, T. L., G. E. Wilde, and K. D. Kofoid. 1997.** Designation of a new greenbug, biotype K, injurious to resistant sorghum. Crop Science 37: 989–991.

115 **Hayes, W. P. 1917.** Studies on the life-history of *Ligyrus gibbosus* DeG. (Coleoptera). J. Econ. Entomol. 10: 253–261.

116 **Helle, W., and M. W. Sabelis. 1985.** Spider mites. Their biology, natural enemies and control, vol. 1A & B. Elsevier, New York.

117 **Henry, T. J. 1988.** Family Anthocoridae Fieber, 1837, the minute pirate bugs, pp.12–28. *In* T. J. Henry and R. C. Froeschner [eds.], Catalog of Heteroptera, or true bugs, of Canada and the continental United States. Leiden, New York.

118 **Higgins, R. A., F. L. Poston, S. M. Welch, D. D. Calvin, B. Droste, J. Pontius, G. Brandsberg, K. Fischer, and R. Wolf. 1986.** European corn borer phenology and management models (user's guide). Kansas State University Cooperative Extension Service, Manhattan.

119 **Higley, L. G., and W. K. Wintersteen. 1992.** A new approach to environmental risk assessment of pesticides as a basis for incorporating environmental costs into economic injury levels. Am. Entomol. 38: 34–39.

120 **Hill, D. S. 1990.** Pests of stored products and their control. CRC Press, Boca Raton, Florida.

121 **Hodek, I. 1973.** Ecology of Coccinellidae. Vol. 54, Series Entomologica. Kluwer Academic, Dordrecht, Netherlands.

122 **Hodgson, C., and C. Aveling. 1988.** Anthocoridae, pp. 279–292. *In* A. K. Minks and P. Harrewijn [eds.], Aphids: their biology, natural enemies and control, vol. B. Elsevier, Amsterdam.

123 **Holland, W. J. 1968.** The moth book. Dover Publications, New York.

124 **Horn, G. H. 1893.** North American Coleoptera. Trans. Am. Entomol. Soc. 20: 111–112.

125 **Hostetter, D. L., and B. Puttler. 1991.** A new broad host spectrum nuclear polyhedrosis virus isolated from a celery looper, *Anagrapha falcifera* (Kirby), (Lepidoptera: Noctuidae). Environ. Entomol. 20: 1480–1488.

126 **Hudon, M. E., E. J. LeRoux, and D. G. Harcourt. 1989.** Seventy years of European corn borer (*Ostrinia nubilalis*) research in North America, pp. 1–44. *In* G. E. Russel [ed.], Biology and population dynamics of invertebrate crop pests. Intercept, Andover, Hampshire.

127 **Hunt, T. N., and J. R. Baker [eds.]. 1982.** Insects and related pests of field crops. North Carolina Agricultural Extension Service AG-271, North Carolina State University, Raleigh.

128 **Illinois Cooperative Extension Service. 1988.** False chinch bugs, p. 143. *In* Insect, Weed & Plant Disease Survey Bulletin No. 13. University of Illinois Cooperative Extension Service, Urbana-Champaign.

129 **Illinois Cooperative Extension Service. 1997.** Illinois Agronomy Handbook, 1997-1998. Cooperative Extenstion Service Circular 1344, University of Illinois College of Agricultural, Consumer and Environmental Sciences, Urbana-Champaign.

130 **Ingram, J. W., E. K. Bynum, R. Mathes, W. E. Haley, and L. J. Charpentier. 1951.** Pests of sugarcane and their control. USDA Circular 878: 18–22, Washington, D. C.

131 **Isenhour, D. J., B. R. Wiseman, and R. C. Layton. 1989.** Enhanced predation by *Orius insidiosus* (Hemiptera: Anthocoridae) on larvae of *Heliothis zea* and *Spodoptera frugiperda* (Lepidoptera: Noctuidae) caused by prey feeding on resistant corn genotypes. Environ. Entomol. 18: 418–422.

132 **Jensen, S. G., L. C. Lane, and D. L. Seifers. 1996.** A new disease of maize and wheat in the high plains. Plant Dis. 80: 1387-1390.

133 **Jepson, L. R., H. H. Keifer, and E. W. Baker. 1975.** Mites injurious to economic plants. University of California Press, Berkeley.

134 **Johnson, W. T., and H. H. Lyon. 1988.** Insects that feed on trees and shrubs. Cornell University Press, Ithaca, New York.

135 **Johnson, T. B., and F. T. Turpin. 1985.** Northern and western corn rootworm (Coleoptera: Chrysomelidae) oviposition in corn as influenced by foxtail populations and tillage systems. J. Econ. Entomol. 78: 57–60.

136 **Kaya, H. K. 1985.** Entomogenous nematodes for insect control in IPM systems, pp. 282–302. *In* M. A. Hoy and D. C. Herzog [eds.], Biological control in agricultural IPM systems. Academic, New York.

137 **Kelleher, J. S., and M. A. Hulme. 1981.** Biological control programs against insects and weeds in Canada 1969–1980. Commonwealth Agricultural Bureaux, Slough, England.

138 **Kelly, E. O. G. 1915.** The southern corn leaf-beetle. USDA Bulletin No. 221, Washington, D. C.

139 **Kirk, V. M. 1959.** Maize billbug control in South Carolina. South Carolina Agricultural Experiment Station Bulletin 452, Clemson.

140 **Knutson, A. E., and F. E. Gilstrap. 1990.** Life tables and population dynamics of the southwestern corn borer (Lepidoptera: Pyralidae) in Texas corn. Environ. Entomol. 19: 684–696.

141 **Knutson, A. E., and F. E. Gilstrap. 1990.** Seasonal occurrence of *Beauveria bassiana* in the southwestern corn borer (Lepidoptera: Pyralidae) in Texas High Plains. J. Kansas Entomol. Soc. 63: 243–251.

142 **Kommedahl, T., and C. E. Windels. 1981.** Root-, stalk-, and ear-infecting *Fusarium* species on corn in the USA, pp. 94–103. *In* P. E. Nelson, T. A. Toussoun, and R. J. Cook [eds.], Fusarium: diseases, biology, and taxonomy. Pennsylvania State University Press, University Park.

143 **Krysan, J. L., and T. A. Miller [eds.]. 1986.** Methods for the study of pest *Diabrotica*. Springer-Verlag, New York.

144 **Krysan, J. L., and R. F. Smith. 1987.** Systematics of the *virgifera* species group of *Diabrotica* (Coleoptera: Chrysomelidae: Galerucinae). Entomography 5: 375–484.

145 **Krysan, J. L., R. F. Smith, T. F. Branson, and P. Guss. 1980.** A new subspecies of *Diabrotica virgifera* (Coleoptera: Chrysomelidae): description, distribution and sexual compatibility. Ann. Entomol. Soc. Am. 73: 123–130.

146 **Landis, D. A., and M. J. Haas. 1992.** Influence of landscape structure on abundance and within-field distribution of European corn borer (Lepidoptera: Pyralidae) larval parasitoids in Michigan. Environ. Entomol. 21: 409–416.

147 **Lasack, P. M., and L. P. Pedigo. 1986.** Movement of stalk borer larvae (Lepidoptera: Noctuidae) from noncrop areas into corn. J. Econ. Entomol. 79: 1697–1702.

148 **Lentz, G. L. 1985.** Occurrence of *Phyllophaga congrua* (LeConte) and *P. implicita* (Horn) (Coleoptera: Scarabaeidae) on soybeans. J. Kansas Entomol. Soc 58: 202–206.

149 **Levine, E. 1983.** Temperature requirements for development of the stalk borer, *Papaipema nebris* (Lepidoptera: Noctuidae). Ann. Entomol. Soc. Am. 76: 892–895.

150 **Levine, E., and H. Oloumi-Sadeghi. 1991.** Management of diabroticite rootworms in corn. Annu. Rev. Entomol. 36: 229–255.

151 **Levine, E., and H. Oloumi-Sadeghi. 1996.** Western corn rootworm (Coleoptera: Chrysomelidae) larval injury to corn grown for seed production following soybeans grown for seed production. J. Econ. Entomol. 89: 1010–1016.

152 **Levine, E., H. Oloumi-Sadeghi, and J. R. Fisher. 1992.** Discovery of multiyear diapause in Illinois and South Dakota northern corn rootworm (Coleoptera: Chrysomelidae) eggs and incidence of the prolonged diapause trait in Illinois. J. Econ. Entomol. 85: 262–267.

153 **Lewis, T. 1973.** Thrips: their biology, ecology, and economic importance. Academic, New York.

154 **Linker, H. M., J. W. Van Duyn, J. R. Anderson, and W. M. Lewis. 1990.** Scouting corn in North Carolina. North Carolina Cooperative Extension Service, AG-399, Raleigh.

155 **Lofgren, C. S. 1986.** The economic importance and control of imported fire ants in the United States, pp. 227–256. *In* S. B. Vinson [ed.], Economic impact and control of social insects. Praeger, New York.

156 **Luginbill, P., and H. R. Painter. 1953.** May beetles of the United States and Canada. USDA Technical Bulletin 1060, Washington, D. C.

157 **Lummus, P. F., J. C. Smith, and N. L. Powell. 1983.** Soil moisture and texture effects on survival of immature southern corn rootworms, *Diabrotica undecimpunctata howardi* Barber (Coleoptera: Chrysomelidae). Environ. Entomol. 12: 1529–1531.

158 **MacKay, M. R. 1962.** Larvae of the North American Tortricinae (Lepidoptera: Tortricidae). Can. Entomol., Supplement 28.

159 **MacVean, C. M. 1987.** Ecology and management of the Mormon cricket, *Anabrus simplex* Haldeman, pp. 116–136. *In* J. L. Capinera [ed.], Integrated pest management on rangeland. Westview Press, Boulder, Colorado.

160 **Malcolm, D. R. 1955.** Biology and control of the timothy mite, *Paratetranychus pratensis* (Banks). Washington Agricultural Experiment Station Technical Bulletin 17, Pullman.

161 **Maramorosch, K., and K. E. Sherman [eds.]. 1985.** Viral insecticides for biological control. Academic Press, Orlando, Florida.

162 **Marsh, H. O. 1910.** Some insects injurious to truck crops. Biologic and economic notes on the yellow-bear caterpillar. USDA, Bureau of Entomology Bulletin 82 V, 59–66, Washington, D. C.

163 **Martignoni, M. E., and P. J. Iwai. 1986.** A catalog of viral diseases of insects, mites, and ticks. USDA, Forest Service General Technical Report PNW-195, Washington, D. C.

164 **Mason, C. E., R. F. Romig, L. E. Wendel, and L. A. Wood. 1994.** Distribution and abundance of larval parasitoids of European corn borer (Lepidoptera: Pyralidae) in the east central United States. Environ. Entomol. 23: 521–531.

165 **Mason, C. E., M. E. Rice, D. D. Calvin, J. W. Van Duyn, W. B. Showers, W. D. Hutchison, J. F. Witkowski, R. A. Higgins, D. W. Onstad, and G. P. Dively. 1996.** European corn borer ecology and management. North Central Regional Extension Publication No. 327, Iowa State University, Ames.

166 **McCauley, G. W., Jr., D. C. Margolies, and J. C. Reese. 1990.** Feeding behavior, fecundity and weight of sorghum- and corn-reared greenbugs on corn. Entomol. Exp. Appl. 55: 183–190.

167 **McLeod, M., D. D. Kopp, and D. K. McBride. 1986.** White grub management. North Dakota State University, Cooperative Extension Service, E-901, Fargo.

168 **McMillian, W. W., N. W. Widstrom, and B. R. Wiseman. 1982.** Pink scavenger caterpillar resistance among selected dent corn hybrids. J. Georgia Entomol. Soc. 17: 93–96.

169 **McPherson, R. M. 1983.** Damage potential of cereal leaf beetles (Coleoptera: Chrysomelidae) in Virginia small grains and corn. J. Econ. Entomol. 76: 1285–1287.

170 **Meinke, L. J., B. D. Siegfried, R. J. Wright, and L. D. Chandler. 1998.** Adult susceptibility of Nebraska western corn rootworm (Coleoptera: Chrysomelidae) populations to selected insecticides. J. Econ. Entomol. 91: 594–600.

171 **Metcalf, Z. P. 1917.** Biological investigations of *Sphenophorus callosus* Olivier. North Carolina Agricultural Experiment Station Technical Bulletin No. 13, Raleigh.

172 **Metcalf, R. L., and W. H. Luckmann [eds.]. 1994.** Introduction to insect pest management, 3rd ed. Wiley, New York.

173 **Metcalf, R. L., and R. A. Metcalf. 1993.** Destructive and useful insects: their habits and control, 5th ed. McGraw-Hill, New York.

174 **Milliken, F. B. 1918.** *Nysius ericae*, the false chinch bug. J. Agric. Res. 13: 571–578, pls. 60–61. [Milliken mistakenly identified *N. raphanus* as *N. ericae*.]

175 **Morrison, W. P., G. B. Cronholm, R. D. Parker, J. W. Stewart, C. D. Patrick, and T. L. Archer. 1991.** Managing insect and mite pests of Texas corn. Texas Agricultural Extension Service Bulletin 1366, College Station.

176 **Munson, G., A. J. Keaster, and J. A. Grundler. 1986.** Corn cutworm control. University of Missouri-Columbia Extension Division, Agricultural Guide 4150, Columbia.

177 **Munson, G., A. J. Keaster, and J. A. Grundler. 1986.** Control of wireworms and other corn soil insects. Univ. of Missouri Columbia, Extension Division, Agricultural Guide 4154, Columbia.

178 **Nault, L. R., and E. D. Ammar. 1989.** Leafhopper and planthopper transmission of plant viruses. Annu. Rev. Entomol. 34: 503–529.

179 **Negron, J. F., and T. J. Riley. 1991.** Seasonal migration and overwintering of the chinch bug (Hemiptera: Lygaeidae) in Louisiana. J. Econ. Entomol. 84: 1681–1685.

180 **Nyvall, R. F. 1981.** Corn lethal necrosis, p. 82; Wheat streak mosaic, pp. 84, 408–409. *In* Field crop disease handbook. AVI Publishing Co., Westport, Connecticut.

181 **Olsen, C. D., K. S. Pike, L. Boydston, and D. Allison. 1993.** Keys for identification of apterous viviparae and immatures of six small grain aphids. J. Econ. Entomol. 86: 137–148.

182 **Ostlie, K. R., W. D. Hutchison, and R. L. Hellmich. 1997.** Bt corn and European corn borer: long-term success through resistance management. North Central Region Ext. Publ. NCR 602. University of Minnesota, St. Paul.

183 **Owens, J. C., C. R. Ward, and G. L. Teetes. 1976.** Current status of spider mites in corn and sorghum, pp. 38–64. *In* Proc. 31st Annual Corn and Sorghum Conference. American Seed Trade Association, Chicago, Illinois.

184 **Pausch, R. D. 1978.** Observations on the biology of the seed corn beetles, *Stenolophus comma* and *Stenolophus lecontei*. Ann. Entomol. Soc. Am. 72: 24–28.

185 **Pausch, R. D., and L. M. Pausch. 1980.** Observations on the biology of the slender seedcorn beetle, *Clivina impressifrons* (Coleoptera: Carabidae). Great Lakes Entomol. 13: 189–194.

186 **Pedigo, L. P. 1989.** Entomology and pest management. Macmillan Publishing Co., New York.

187 **Pfadt, R. E. 1988.** Field guide to common western grasshoppers. USDA-APHIS, Wyoming Agricultural Experiment Station Bulletin 912, Laramie.

188 **Phillips, W. J. 1914.** Corn-leaf blotch miner. J. Agric. Res. 2: 15–31.

189 **Phillips, W. J., and H. Fox. 1933.** The rough-headed corn stalk beetle in the southern states and its control. USDA Farmers Bull. 875, Washington, D. C.

190 **Pioneer Hi-Bred International, Inc. 1987.** Corn management/diagnostic guide. Pioneer Hi-Bred International, Johnston, Iowa.

191 **Poinar, O., Jr., and G. M. Thomas. 1984.** Laboratory guide to insect pathogens and parasites. Plenum Press, New York.

192 **Prokrym, D. R., D. A. Andow, J. A. Ciborowski, and D. D. Sreenivasam. 1992.** Suppression of *Ostrinia nubilalis* by *Trichogramma nubilale* in sweet corn. Entomol. Exp. Appl. 64: 73–85.

193 **Puterka, G. J., J. E. Slosser, and J. P. Price. 1985.** Parasites of *Heliothis* species (Lepidoptera: Noctuidae): parasitism and seasonal occurrence for host crops in the Texas rolling plains. Environ. Entomol. 14: 441–446.

194 **Ratcliffe, B. C. 1991.** The scarab beetles of Nebraska. Bulletin of the University of Nebraska State Museum No. 12, Lincoln.

195 **Raulston, J. R., S. D. Pair, J. Loera, A. N. Sparks, W. W. Wolf, J. K. Westbrook, G. P. Fitt, and C. E. Rogers. 1992.** *Helicoverpa zea* (Lepidoptera: Noctuidae) pupa production in fruiting corn in northeast Mexico and South Texas. Environ. Entomol. 21: 1393–1397.

196 **Redak, R. A., J. L. Capinera, and C. D. Bonham. 1992.** Effects of sagebrush removal and herbivory by Mormon crickets (Orthoptera: Tettigoniidae) on understory plant biomass and cover. Environ. Entomol. 21: 94–102.

197 **Reid, C. D. 1991.** Ability of *Orius insidiosus* (Hemiptera: Anthocoridae) to search for, find, and attack European corn borer and corn earworm eggs on corn. J. Econ. Entomol. 84: 83–86.

198 **Reid, W. J., Jr. 1940.** Biology of the seedcorn maggot in the coastal plain of the south Atlantic states. USDA Technical Bulletin 723, Washington, D. C.

199 **Rice, M. E. 1993.** Leafrollers in corn. Integrated Crop Management IC-466: 106. Iowa State University, Ames.

200 **Rice, M. E. 1994.** Damage assessment of the annual white grub, *Cyclocephala lurida* (Coleoptera: Scarabaeidae), in corn and soybean. J. Econ. Entomol. 87: 220–222.

201 **Riley, T. J., and A. J. Keaster. 1981.** A pictorial field key to wireworms attacking corn in the Midwest. University of Missouri Publication MP 517, Columbia.

202 **Rings, R. W. 1977.** An illustrated field key to common cutworm, armyworm, and looper moths in the north central states. Ohio Agricultural Research and Development Center Research Circular 227, Wooster.

203 **Rings, R. W., and F. J. Arnold. 1974.** An annotated bibliography of the glassy cutworm, *Crymodes devastator* (Brace). Ohio Agricultural Research and Development Center Research Circular 199, Wooster.

204 **Rings, R. W., and F. J. Arnold. 1975.** An annotated bibliography of the bristly cutworm, *Lacinipolia renigera* (Stephens). Ohio Agricultural Research and Development Center Research Circular 201, Wooster.

205 **Rings, R. W., and B. A. Johnson. 1976.** An annotated bibliography of the sandhill cutworm, *Euxoa detersa* (Walker). Ohio Agricultural Research and Development Center Research Circular 215, Wooster.

206 **Rings, R. W., and G. J. Musick. 1976.** A pictorial field key to the armyworms and cutworms attacking corn in the north central states. Ohio Agricultural Research and Development Center Research Circular 221, Wooster.

207 **Rings, R. W., F. J. Arnold, A. J. Keaster, and G. J. Musick. 1974.** A worldwide, annotated bibilography of the black cutworm, *Agrotis ipsilon* (Hufnagel). Ohio Agricultural Research and Development Center Research Circular 198, Wooster.

208 **Rings, R. W., B. A. Baughman, and F. J. Arnold. 1974.** An annotated bibliography of the bronzed cutworm, *Nephelodes minians* Guenee. Ohio Agricultural Research and Development Center Research Circular 200: 1-37, Wooster.

209 **Rings, R. W., B. A. Baughman, and F. J. Arnold. 1975.** An annotated bibliography of the dingy cutworm complex *Feltia ducens* Walker and *Feltia subgothica* (Haworth). Ohio Agricultural Research and Development Center Research Circular 202, Wooster.

210 **Rings, R. W., B. A. Johnson, and F. J. Arnold. 1975.** An annotated bibliography of the dark-sided cutworm *Euxoa messoria* (Harris). Ohio Agricultural Research and Development Center Research Circular 205, Wooster.

211 **Rings, R. W., B. A. Johnson, and F. J. Arnold. 1976.** An annotated bibliography of the clay-backed cutworm, *Agrotis gladiaria* (Morrison). Ohio Agricultural Research and Development Center Research Circular 216, Wooster.

212 **Rings, R. W., B. A. Johnson, and F. J. Arnold. 1976.** An annotated bibliography of the dusky cutworm, *Agrotis venerabilis* Walkder. Ohio Agricultural Research and Development Center Research Circular 217, Wooster.

213 **Ritchie, S. W., J. J. Hanway, and G .O. Benson. 1992.** How a corn plant develops. Special Report No. 48, Iowa State University, Ames.

214 **Roberts, R. A. 1952.** Whitefringed beetle, pp. 608–611. *In* Insects. 1952 Yearbook of Agriculture, USDA, Washington, D. C.

215 **Ruberson, J. R., M. J. Tauber, and C. A. Tauber. 1986.** Plant feeding by *Podisus maculiventris* (Heteroptera: Pentatomidae): effect on survival, development and preoviposition period. Environ. Entomol. 15: 894–897.

216 **Sanford, J. W., and W. H. Luckmann. 1963.** Observations on the biology and control of the dusky sap beetle in Illinois. Proc. North Central Branch Entomol. Soc. Am. 18: 39–43.

217 **Sedlacek, J. D., and L. H. Townsend. 1988.** Impact of *Euschistus servus* and *E. variolarius* (Heteroptera: Pentatomidae) feeding on early growth stages of corn. J. Econ. Entomol. 81: 840–844.

218 **Segers, J. C., J. H. Benedict, and M. F. Treacy. 1984.** Natural parasitism of *Heliothis* spp. eggs on cotton and corn in the lower Gulf coast of Texas. Southwestern Entomol. 9: 245–248.

219 **Shetlar, D. J. 1982.** Earwigs, pp. 167–175. *In* A. Mallis [ed.], Handbook of pest control, 6th ed. Franzak & Foster Co., Cleveland, Ohio.

220 **Showers, W. B., L. Von Kaster, and P. Mulder. 1983.** Corn seedling growth stage and black cutworm (Lepidoptera: Noctuidae) damage (*Agrotis ipsilon*). Environ. Entomol. 12: 241–244.

221 **Showers, W. B., L. V. Kaster, T. W. Sappington, P. G. Mulder, and F. Whitford. 1985.** Development and behavior of black cutworm (Lepidoptera: Noctuidae) populations before and after corn emergence. J. Econ. Entomol. 78: 588–594.

222 **Showers, W. B., A. J. Keaster, J. R. Raulston, W. H. Hendrix III, M. E. Derrick, M. D. McCorcle, J. F. Robinson, M. O. Way, M. J. Wallendorf, and J. L. Goodenough. 1993.** Mechanism of southward migration of a noctuid moth [*Agrotis ipsilon* (Hufnagel)]: A complete migrant. Ecology 74: 2303–2314.

223 **Shufran, K. 1992.** Genotypic and phenotypic variability of Kansas greenbug populations. Ph.D. dissertation. Kansas State University, Manhattan.

224 **Shurtleff, M. C. [ed.]. 1980.** Compendium of corn diseases. APS Press, St. Paul, Minnesota.

225 **Smith, M.R. 1965.** House infesting ants of the eastern United States. USDA Technical Bulletin 1326: 43–45 and 80–81, Washington, D. C.

226 **Smith, D. R., and D. G. White. 1988.** Diseases of corn, pp. 687–766. *In* G. F. Sprague and J. W. Dudley [eds.], Corn and corn improvement. Agronomy Monograph No. 18, 3rd ed. American Society of Agronomy, Crop Science Society of America, and Soil Science Society of America, Madison, Wisconsin.

227 **Smith, M. T., G. Wilde, and T. Mize. 1981.** Chinch bug: damage and effects of host plant and photoperiod. Environ. Entomol. 10: 122–124.

228 **Sornenson, C. J., and H. F. Thornley. 1941.** The pale western cutworm (*Agrotis orthogonia* Morrison) in Utah. Utah Agricultural Experiment Station Bulletin 297, Ogden.

229 **Sparks, A.N., and E. R. Mitchell. 1979.** Economic thresholds of *Heliothis* species on corn. *In* W. L. Sterling, Chair, Economic thresholds and sampling of *Heliothis* species on cotton, corn, soybeans and other host plants. Southern Cooperative Series Bulletin 231. Texas Agricultural Experiment Station, College Station.

230 **Spencer, K. A. 1973.** Agromyzidae (Diptera) of economic importance. Vol. 9, Series Entomologica. Dr. W. Junk B. V., The Hague, The Netherlands.

231 **Spike, B. P., G. E. Wilde, T. W. Mize, R. J. Wright, and S. D. Danielson. 1994.** Bibliography of the chinch bug, *Blissus leucopterus leucopterus* (Say) (Heteroptera: Lygaeidae) since 1888. J. Kansas Entomol. Soc. 67: 116–125.

232 **Sprague, G. F., and J. W. Dudley [eds.]. 1988.** Corn and corn improvement. Agronomy Monograph No. 18, 3rd ed. American Society of Agronomy, Crop Science Society of America, and Soil Science Society of America, Madison, Wisconsin.

233 **Starks, K. J., H. C. Cox, W. W. McMillian, and R. L. Burton. 1966.** Damage to corn by the pink scavenger caterpillar and its relationship to corn earworm and rice weevil damage. J. Econ. Entomol. 59: 931–934.

234 **Stehr, F. W. 1987.** Immature insects. Kendall/Hunt Publishing, Dubuque, Iowa.

235 **Stern, V. M., R. E. Smith, R. van den Bosch, and K. S. Hagen. 1959.** The integrated control concept. Hilgardia 29: 81–101.

236 **Stinner, B. R., and G. J. House. 1990.** Arthropods and other invertebrates in conservation-tillage agriculture. Annu. Rev. Entomol. 35: 299–318.

237 **Stork, N. E. 1990.** The role of ground beetles in ecological and environmental studies. Intercept, Andover, Hampshire.

238 **Straub, R. W. 1984.** Maize dwarf mosaic virus: symptomatology and yield reactions of susceptible and resistant sweet corns. Environ. Entomol. 13: 318–323.

239 **Summers, T. E., E. G. King, D. F. Martin, and R. D. Jackson. 1976.** Biological control of *Diatraea saccharalis* [Lep.: Pyralidae] in Florida by periodic releases of *Lixophaga diatraeae* [Dipt.: Tachninidae]. Entomophaga 21: 359–366.

240 **Sutter, G. R., and D. R. Lance. 1991.** New strategies for reducing insecticide use in the Corn Belt, pp. 231–249. *In* Sustainable agriculture research and education in the field: a proceedings. National Research Council, National Academy of Sciences, Washington, D. C.

241 **Tanada, Y., and H. K. Kaya. 1992.** Insect pathology. Academic, New York.

242 **Tashiro, H. 1987.** Turfgrass insects of the United States and Canada. Cornell University Press, Ithaca, New York.

243 **Throne, J. E. 1980.** Bibliography of seed maggots, *Hylemya platura* and *H. florilega* (Diptera: Anthomyiidae). New York State Agricultural Experiment Station, Geneva Special Report No. 37, Geneva.

244 **Tippins, H. H. 1982.** A review of information on the lesser cornstalk borer, *Elasmopalpus lignosellus* (Zeller). University of Georgia Experiment Station Special Publication 17, Athens.

245 **Tollefson, J. J., and D. D. Calvin. 1994.** Sampling arthropod pests in field corn, pp. 433–473. *In* L. Pedigo and D. Buntin [eds.], Handbook of sampling methods for arthropod pests in agriculture. CRC Press, Boca Raton, Florida.

246 **Townes, H. 1972.** Ichneumonidae as biological control agents, pp. 235–248. *In* Proceedings of the Tall Timbers Conference on Ecological Animal Control by Habitat Management 1971. Tall Timbers Research Station, Tallahassee, Florida.

247 **United States Department of Agriculture. 1972.** Controlling whitefringed beetles. Leaflet 550, Washington, D. C.

248 **Van Duyn, J. W. 1985.** Pest management in field corn. *In* J. R. Anderson and M. G. Wagger [eds.], Corn production systems in North Carolina. North Carolina Agricultural Extension Service Publication AG-347, Raleigh.

249 **Vea, E. V., D. R. Webb, and C. J. Eckenrode. 1975.** Seedcorn maggot injury. New York Food and Life Sciece Bulletin 55, Ithaca.

250 **Vinson, S. B., and L. Greenburg. 1986.** The biology, physiology, and ecology of imported fire ants, pp. 193–226. *In* S. B. Vinson [ed.], Economic impact and control of social insects. Praeger, New York.

251 **Wajnberg, E., and S. B. Vinson [eds.]. 1991.** *Trichogramma* and other egg parasitoids. 3rd International Symposium, San Antonio, Texas, September 1990. INRA, Paris.

252 **Wakeland, C. 1959.** Mormon crickets in North America. USDA, Agricultural Research Service Technical Bulletin 1202, Washington, D. C.

253 **Walkden, H. H. 1950.** Cutworms, armyworms and related species attacking cereal and forage crops in the Central Great Plains. USDA Circular 849, Washington, D. C.

254 **Way, M. J., and M. E. Cammell. 1981.** Effects of weeds and weed control on invertebrate pest ecology, pp. 443–458. *In* J.M. Thresh [ed.], Pests, pathogens, and vegetation: the role of weeds and wild plants in the ecology of crop pests and diseases. Pitman, Boston, Massachusetts.

255 **Wedberg, J. L., J. B. Campbell, and T. J. Helms. 1975.** Bionomics of insects associated with corn in the Nebraska Sandhills. University of Nebraska Research Bulletin 274, Lincoln.

256 **Wellso, S. G., and R. P. Hoxie. 1988.** Biology of *Oulema*, pp. 497–511. *In* P. Jolivet, E. Petitpierre and T. Hsiao [eds.], Biology of Chrysomelidae. Vol. 42, Series Entomologica. Kluwer Academic Publishers, Boston, Massachusetts.

257 **Wildermuth, V. L., and E. V. Walter. 1932.** Biology and control of the corn leaf aphid with special reference to the southwestern states. USDA Technical Bulletin 306, Washington, D. C.

258 **Willson, H. R., and J. B. Eisley. 1992.** Effects of tillage and prior crop on the incidence of five key pests on Ohio corn. J. Econ. Entomol. 85: 853–859.

259 **Wiseman, B. R., and F. M. Davis. 1990.** Plant resistance to insects attacking corn and grain sorghum. Florida Entomol. 73: 446–458.

260 **Wiseman, B. R., and W. P. Morrison. 1981.** Components for management of field corn and grain sorghum insects and mites in the United States. USDA-ARS ARM-S-18. Agricultural Research Service (Southern Region), USDA, New Orleans, Louisiana.

261 **Woodring, J. L., and H. K. Kaya. 1988.** Steinernematid and heterorhabditid nematodes: a handbook of techniques. Arkansas Agricultural Experiment Station, Southern Cooperative Service Bulletin 311, Fayetteville.

262 **Wright, E. J., and J. E. Laing. 1980.** Numerical response of coccinellids to aphids in corn in southern Ontario. Can. Entomol. 112: 977–988.

Glossary

abdomen. The third or posteriormost of the three major body divisions of an insect, or the second or posteriormost of the two major body divisions of spiders and mites.

adventitious roots. Roots that originate from nodes on the stem.

aflatoxin. Any of several carcinogenic mycotoxins that are produced by molds, especially in stored agricultural crops.

alate. The winged form of an aphid.

anal plate. A hardened "plate" on the dorsal surface of the last abdominal segment.

anthers. The parts of male flowers that produce pollen; located on tassels.

apterous. Wingless form of an insect.

augmentation. A method of biological control that enhances effectiveness of a biological control agent by mass release, genetic improvement, or use of semiochemicals to alter behavior.

baculovirus. Virus in the family Baculoviridae, which are rod-shaped, double-stranded DNA viruses.

basement membrane. A noncellular membrane underlying the epidermal cells of the body wall.

biological control. The actions of parasites, parasitoids, predators, and pathogens that lower host population densities; and the study and use of parasites, parasitoids, predators, and pathogens for the regulation of host population densities.

biotype. An individual that is different from the rest of the species by some criteria other than morphology, for example the ability to injure a plant that is resistant to the rest of the population.

bivoltine. Having two generations per year.

black layer. A black abscission layer near the tip of a corn kernel that indicates that the transport of photosynthate into the kernel has stopped; connotes maturity in corn.

caterpillar. Usually refers to the larva of a moth or butterfly.

chlorosis. Yellowing or blanching of green plants.

chorion. The outermost protective layer around an insect egg; the "shell."

coleoptile. The sheath enclosing the apical meristem and leaf primordia of the grass embryo; often interpreted as the first leaf.

collar. The junction between the leaf blade and leaf sheath in grasses such as corn.

complete metamorphosis. Metamorphosis of an insect that includes four stages of development: egg, larva, pupa, adult.

conservation. A method of biological control based upon environmental manipulation to enhance the effectiveness of a biological agent.

conidium. An asexual fungal spore produced on a conidiophore.

Corn Belt. A multistate region within the midwestern United States where corn is a staple crop.

cornicles. Erect tubules located on the end of the abdomen of aphids that secrete a defensive liquid.

crown. Region at the base of the stem from which tillers arise.

cytoplasmic polyhedrosis virus. Virus in the family Reoviridae, which are icosahedral, double-stranded RNA viruses.

"dead heart." Wilting, discoloration, and death of whorl leaves when stem feeding (boring) activity of insects cuts off water and nutrient flow.

densonucleosis virus. Virus in the family Parvoviridae, which are small, isometric, single-stranded DNA viruses; also called a densovirus.

deutonymph. The third instar of a mite.

diapause. A period of arrested development and reduced metabolic rate during which growth, differentiation, and metamorphosis cease; a period of dormancy not immediately referable to adverse environmental conditions.

ear. The female flowering structure of the corn plant.

ear shank. The part of the corn plant that attaches the ear to the stalk.

ear shoot. The sprout of a corn plant that develops into the ear.

economic injury level. The pest density or level of injury that will result in a yield loss equal to the cost of controlling the pest.

economic threshold. The pest density at which management action should be taken to prevent an increasing pest populaton from reaching the economic injury level.

ectoparasitoid. A parasitoid in which immature stages feed and develop outside the host.

elytra (singular **elytron**). Thickened, leathery, or horny front wings, primarily in Coleoptera (beetles) and Dermaptera (earwigs).

endoparasitoid. A parasitoid in which immature stages feed and develop inside the host.

endotoxin. A poisonous substance present in bacteria but separable from the cell body only on its disintegration.

entomogenous. Microorganisms growing in or on the bodies of insects; refers to a parasitic relationship.

entomopathogen. Organism that is pathogenic to arthropods.

entomopoxvirus. Virus in the family Poxviridae, which are ovoid, double-stranded DNA viruses with a proteinaceous occlusion body.

epidermis. The cellular layer of the body wall, which secretes the cuticle.

epizootic. Outbreak of a disease that suddenly and temporarily affects an animal population.

estivation. Dormancy during a warm or dry season.

eucaryote. An organism composed of one or more cells with visibly evident nuclei.

feces. Excrement, the material passed from the alimentary tract through the anus.

femur. The third leg segment, located between the trochanter and the tibia.

frass. Plant fragments made by a plant-boring insect, usually mixed with excrement.

granulosis virus. An occluded baculovirus in which the occlusion body is round to ellipsoidal, 0.1–1μ in diameter.

growing point. A part of the plant body at which cell division is localized, generally terminal and composed of meristematic cells.

grub. Thick-bodied larva with a well developed head and thoracic legs, without abdominal prolegs, and usually sluggish. Term used to describe some beetle larvae.

heat unit. An accumulation of degrees above some threshold temperature for a 24-hour period; a measure of physiological time for cool-blooded organisms, like insects; also called degree-day.

hemocoel. A body cavity filled with blood.

holometabolous. The higher insects that pass through a complete metamorphosis with distinct egg, larval, pupal, and adult stages.

honeydew. A liquid rich in sugar excreted from the anus of an aphid.

hyperparasite. An insect that parasitizes another parasitic insect.

importation or classical biological control. A method of biological control that introduces exotic natural enemies.

infection. The introduction or entry of a pathogenic microorganism into a susceptible host, resulting in the presence of the microorganism within the body of the host, whether or not this causes detectable pathological effects.

instar. The stage of an insect between successive molts, the first instar being the stage between hatching and the first molt.

intercropping. The growing together of two or more crops in the same field at the same time.

iridescent virus. Virus in the family Iridoviridae, which are large, isometric DNA viruses.

labial palps. A pair of small feelerlike structures arising from the labium (part of the mouthparts of an insect).

larva. The immature stage between egg and pupa in insects with complete metamorphosis.

leaf sheath. The extension of the leaf base sheathing the stem.

lumen. Gut lumen; the cavity inside the gut.

maggot. A legless larva without a well developed head capsule. A term used to described some fly larvae.

mandibles. Part of the mouthparts of insects, usually used to chew food, but sometimes modified into structures for piercing or sucking.

meristem. The undifferentiated plant tissue from which new cells arise.

mesocotyl. The internode between the seed and the coleoptile in grass plants.

metamorphosis. A change in form during development.

migration. Any cyclical movement (usually annual) that occurs during the life history an an animal at definite intervals and always includes a return trip from where they began.

monocot. A subdivision of flowering plants whose members possess one embryonic seed leaf, or cotyledon; includes corn.

multivoltine. Having many generations per year.

mycotoxicosis. Disease condition caused by the action of a toxin produced by a fungus.

natural control. The actions of abiotic factors (such as weather) and biotic factors (such as natural enemies) that lower population densities of an organism.

necrosis. Usually localized death of living tissue.

node. Knob or joint of a stem at which leaves arise.

nonoccluded baculovirus. A baculovirus for which the virion capsids are not embedded in a protienaceous crystal.

nuclear polyhedrosis virus. An occluded baculovirus in which the occlusion body is polyhedral to cuboidal, 1–15μ in diameter.

nymph. The immature stage in insects with incomplete metamorphosis.

occluded baculovirus. A baculovirus for which the virion capsids are embedded in a large proteinaceous crystal; the crystal also is referred to as an occlusion body. These viruses are the granulosis viruses and the nuclear polyhedrosis viruses.

oviposit. To lay or deposit eggs.

ovipositor. The tubular structure through which female insects deposit their eggs. In wasps it is the "stinger."

parasitoid. A species that parasitizes another species, in which a relatively small number of immature stages slowly consume the host, and adult stages actively search for hosts; also called a protelean parasite. These include Tachinidae and parasitic Hymenoptera.

parthenogenesis. Asexual reproduction; eggs undergo full development without being fertilized.

pathogen. A specific cause of disease. A microorganism capable of producing disease under normal conditions of host resistance and rarely living in close association with the host without producing disease. Any microorganism, virus, substance, or factor causing disease.

pedicel. The "waist" of an ant, made up of either one or two segments, or the second segment of an antenna from the base outward.

pericarp. The fruit wall, which develops from the mature ovary wall.

pollen. Microspores produced by the male flower; usually appears as a fine dust.

prolegs. Fleshy, nonsegmented abdominal walking appendages of some insect larvae.

pronotum. The dorsal "plate" of the prothorax.

prothoracic shield. Another term for pronotum.

protonymph. The second instar of a mite.

pubescence. Short, fine hairs.

pupa. The stage between the larva and the adult in insects with complete metamorphosis; a nonfeeding and usually an inactive stage.

puparium. A case formed by the hardening of the last larval skin, in which the pupa is formed (in Diptera).

raster. Complex of definitely arranged bare areas, hairs, and spines on the ventral surface of the last abdominal segment, in front of the anus.

roost. Dense vegetation where birds sleep at night.

saprophytic. Living on dead or decaying matter.

septicemia. A morbid condition caused by the multiplication of microorganisms in the blood.

setae (singluar **seta**). Bristles.

silks. Long slender tubes produced by ovules on the corn ear; the silks receive pollen and conduct the genetic contents of the pollen grain to the female flower on the cob.

spiracle. An external opening of the tracheal system; a breathing pore.

spores. Reproductive body of the lower plants.

sporozoite. A usually motile infective form of some sporozoans that is a product of sporogony and initiates an asexual cycle in the new host.

stadium (plural **stadia**). The period between molts of a devloping arthropod.

systemic insecticide. An insecticide that is absorbed by plants and is translocated (moved) to other tissues.

tarsal claw. Structure found at the end of the tarsus.

tarsi (singular **tarsus**). Leg segments immediately beyond the tibia, sometimes consistening of one or more segments or subdivisions.

tassel. The male flowering structure of the corn plant from which pollen is shed.

thorax. The body region of an insect between the head and the abdomen that bears the legs and wings.

tibia. The fourth segment of the leg, between the femur and the tarsus.

transgenic organism. An organism containing certain genes from another species; produced, for example, by injecting foreign DNA into the nuclei of egg cells or early embryos.

tubercle. A small knotlike or rounded protuberance.

univoltine. Having one generation per year.

urogomphi. Fixed or movable cercus-like processes on the last segment of a beetle larva.

vector. An organism, such as an insect, that transmits a pathogen.

whorl. A circle of leaves in a corn plant.

worker ant. A member of a caste of nonreproductive adult ants.

Sources of Local Information

Department of Entomology
301 Funchess Hall
Auburn University
Auburn University, AL 36849-5413
(205) 844-5006
FAX (205) 844-5005
http://www.ag.auburn.edu/dept/ent/ent.html

Department of Entomology
University of Arizona
Tucson, AZ 85721
(520) 621-1151
FAX (520) 621-1150

Department of Entomology
University of Arkansas
Fayetteville, AR 72701
(501) 575-2451
FAX (501) 575-2452
E-MAIL wyearia@comp.uark.edu

Department of Environmental Science,
Policy, and Management
Division of Insect Biology
University of California
Berkeley, CA 94720
(510) 642-1603
FAX (510) 642-7428

Department of Entomology
University of California
Davis, CA 95616-8584
(916) 752-0475
FAX (916) 752-1537

Department of Entomology
University of California
Riverside, CA 92521
(909) 787-3718
FAX (909) 787-3086
insects@ucrac1.ucr.edu

Department of Bioagricultural Sciences
and Pest Management
Colorado State University
Fort Collins, CO 80523-1177
(970) 491-5261
FAX (970) 491-3862
bspm@lamar.colostate.edu
http://www.colostate.edu/depts/bspm/index.html

Department of Ecology & Evolutionary Biology
University of Connecticut
U-43
75 N Eagleville Rd
Storrs, CT 06269-3043
(860) 486-4322/4319
FAX (860) 486-6364

Department of Entomology & Applied Ecology
University of Delaware
Newark, DE 19717-1303
(302) 831-2526
FAX (302) 831-3651
jhough@udel.edu

Department of Entomology
National Museum of Natural History, Rm. W308
Smithsonian Institution
Washington, DC 20560-0105
(202) 357-2078
FAX (202) 786-2894
hevelg@nmnh.si.edu

Department of Entomology & Nematology
P.O. Box 110620
University of Florida
Gainesville, FL 32611
(904) 392-1901
FAX (904) 392-0190

Center for Studies in Entomology
Florida A & M University
Tallahassee, FL 32307-4100
(904) 599-3912
FAX (904) 561-2221

Department of Entomology
413 Biological Sciences Bldg.
University of Georgia
Athens, GA 30602-2603
(706) 542-2023
FAX (706) 542-2279
rnoblet@bugs.ent.uga.edu

Department of Entomology
Coastal Plain Experiment Station
University of Georgia
Tifton, GA 31793
(912) 386-3374
FAX (312) 386-3086
entotif@tifton.cpes.peachnet.edu

Department of Entomology
3050 Maile Way, Gilmore 310
University of Hawaii at Manoa
Honolulu, HI 96822
(808) 956-7076
FAX (808) 956-2428
ta_entol@avax.ctahr.hawaii.edu

Division of Entomology
University of Idaho
Moscow, ID 83844-2339
(208) 885-7543
FAX (208) 885-7760
djohnson@uidaho.edu

Illinois Natural History Survey
607 East Peabody
Champaign, IL 61820
(217) 244-2149

Center for Economic Entomology
Illinois Natural History Survey
607 East Peabody Drive
Champaign, IL 61820
(217) 333-6656
FAX (217) 333-4949
armbrust@uiuc.edu

Center for Biodiversity
Illinois Natural History Survey
607 E. Peabody Drive
Champaign, IL 61820
(217) 333-6846
FAX (217) 333-4949
glevin@mail.inhs.uiuc.edu

Department of Entomology
University of Illinois
320 Morrill Hall
505 S. Goodwin Ave.
Urbana, IL 61801
(217) 333-2910
FAX (217) 244-3499
entwork@life.uiuc.edu
http://www.life.uiuc.edu/entomology/home.html

Department of Entomology
Purdue University
1158 Entomology Hall
West Lafayette, IN 47907-1158
(765) 494-4554
FAX (765) 494-0535
chris_oseto@entm.purdue.edu

Department of Entomology
411 Science II
Iowa State University
Ames, IA 50011-3222
(515) 294-7400
FAX (515) 294-5957
tcbaker@iastate.edu

Department of Entomology
University of Kansas
Lawrence, KS 66045
(785) 864-7366
FAX (785) 864-5321
greenfie@kuhub.cc.ukans.edu

Department of Entomology
123 Waters Hall
Kansas State University
Manhattan, KS 66506-4004
(785) 531-6154
FAX (785) 532-6232
entomology@ksu.edu

Department of Entomology
University of Kentucky
Lexington, KY 40546-0091
(606) 257-7450
FAX (606) 323-1120
bpass@ca.uky.edu

Department of Entomology
Louisiana State University
402 Life Sciences Bldg.
Baton Rouge, LA 70803-1710
(504) 388-1634
FAX (504) 388-1643
fguillo@unix1.sncc.lsu.edu

Systematic Entomology Laboratory
Bldg. 046, BARC-West
Agricultural Research Service, USDA
10300 Baltimore Avenue
Beltsville, MD 20705
(301) 504-5183
FAX (301) 504-6482

Department of Entomology
University of Maryland
4112 Plant Science Bldg.
College Park, MD 20742-4454
(301) 405-3911
FAX (301) 314-9290

Department of Entomology
Fernald Hall
University of Massachusetts
Amherst, MA 01003
(413) 545-2283
FAX (413) 545-2115
entomology@ent.umass.edu

Department of Entomology
Michigan State University
243 Natural Science Building
East Lansing, MI 48824
(517) 355-4662
FAX (517) 353-4354
kolpj@pilot.msu.edu
http://www.ent.msu.edu/

Department of Entomology
219 Hodson Hall
1980 Folwell Ave
University of Minnesota
St. Paul, MN 55108-6125
(612) 624-3278
FAX (612) 625-5299
entodept@gold.tc.umn.edu

Department of Entomology & Plant Pathology
Box 9775
Mississippi State University
Mississippi State, MS 39762-9775
(601) 325-2085
FAX (601) 325-8837
chc2@ra.msstate.edu

Department of Entomology
Montana State University
Bozeman, MT 59717
(406) 994-3860
FAX (406) 994-6029
gdj@montana.edu

Department of Entomology
University of Nebraska
Lincoln, NE 68583-0816
(402) 472-2123
FAX (402) 472-4687
squisenb@unlinfo.unl.edu

Rutgers, The State Univ of New Jersey
Blake Hall
93 Lipman Drive
New Brunswick, New Jersey 08901-8524
(908) 932-9459
FAX (908) 932-7229
hamilton@aesop.rutgers.edu

Department of Entomology, Plant Pathology
& Weed Science
New Mexico State University
Box 30003, Dept 3BE
Las Cruces, NM 88003-0003
(505) 646-3225
FAX (505) 646-8087
gkinzer@murphie.nmsu.edu

Sources of Local Information

Department of Entomology-Geneva
Cornell University
NYS Agricultural Experiment Station
Geneva, NY 14456
(315) 781-2323
FAX (315) 787-2326

Department of Entomology
Cornell University
Ithaca, NY 14853
(607) 255-3253
FAX (607) 255-0939
dar11@cornell.edu

Entomology Concentration Leader
133 Illick Hall
SUNY College of Environment Science and Forestry
Syracuse, NY 13210-2788
(315) 470-6742
FAX (315) 470-6934
tnakatsu@syr.edu

Department of Entomology
North Carolina State University
Raleigh, NC 27695-7613
(919) 515-8888
FAX (919) 515-7746
james_harper@ncsu.edu

Department of Entomology
North Dakota State University
202 Hultz Hall
Box 5346, University Station
Fargo, ND 58105-5446
(701) 237-7582
FAX (701) 237-8557

Department of Entomology
Ohio State University
Columbus, OH 43210
(614) 292-8209
FAX (614) 292-2180
montano.3@osu.edu

Department of Entomology & Plant Pathology
127 Noble Research Center
Oklahoma State University
Stillwater, OK 74078
(405) 774-5527
FAX (405) 744-6039
rew0675@okway.okstate.edu

Department of Entomology
Cordley Hall, Room 2046
Oregon State University
Corvallis, OR 97331-2709
(503) 737-4733
FAX (503) 737-3643
ent-off@bcc.orst.edu

Department of Entomology
Pennsylvania State University
University Park, PA 16802
(814) 865-1895
FAX (814) 865-3048
jffz@psu.edu

Crop Protection Department
P.O. Box 9030
University of Puerto Rico
Mayaquez, PR 00681-9030
(809) 265-3859
FAX (809) 265-0860
c_cruz@rumac.upr.clu.edu

Department of Plant Sciences
University of Rhode Island
Kingston, RI 02881
(401) 874-2924
FAX (401) 874-5296
casa@uriacc.uri.edu

Department of Entomology
114 Long Hall, Box 340365
Clemson Univ
Clemson, SC 29634
(864) 656-3111
FAX (864) 656-5065
dpento@clemson.edu

Department of Plant Science
South Dakota State University
Box 2207-A
Brookings, SD 57007
(605) 688-5123
FAX (605) 688-4602
gallend@mg.sdstate.edu

Department of Entomology and Plant Pathology
University of Tennessee
Knoxville, TN 39701
(615) 947-7135

Department of Entomology
Texas A & M University
College Station, TX 77843-2475
(409) 845-2516
FAX (409) 845-6305
entomain@tamu.edu
http://entowww.tamu.edu

Department of Plant and Soil Science
Box 42122
Texas Tech University
Lubbock, TX 79409-2122
(806) 742-2838
FAX (806) 742-0775
zasto@ttacs.ttu.edu

Insect Biology Division
Department of Biology
Utah State University
Logan, UT 84322-5305
(435) 797-2515
FAX (435) 797-1575
entpath@ext.usu.edu

Department of Zoology
Brigham Young University
Provo, UT 84602
(801) 378-5492
FAX (801) 378-3733
richard_baumann@byu.edu

Department of Biology
257 South 1400 East, Room 201
University of Utah
Salt Lake City, UT 84112-0840
(801) 581-6517
FAX (801) 581-4668
lynellgardner@bioscience.utah.edu

Plant & Soil Science Department
Hills Bldg.
University of Vermont
Burlington, VT 05405-0082
(802) 656-2630
FAX (802) 656-4656

Department of Entomology
Virginia Tech
Blacksburg, VA 24061-0319
(540) 231-6341
FAX (540) 231-9131
Mail Code 0319
tmack@vt.edu

Department of Entomology
FSHN Bldg. 166
Washington State University
Pullman, WA 99164-6382
(509) 335-5505
FAX (509) 335-1009
entom@wsu.edu

Division of Plant and Soil Sciences
West Virginia University
P.O. Box 6108
Morgantown, WV 26506-6108
(304) 293-4817 (Division)
(304) 293-6023 (Entomology)
FAX (304) 293-3740

Department of Renewable Resources
University of Wyoming
P.O. Box 3354 University Station
Laramie, WY 82071-3103
(307) 766-3103
FAX (307) 766-3379

Sources of Local Information

Department of Biological Sciences
CW 4-05 Biological Sciences Bldg.
University of Alberta
Edmonton, Alberta
Canada T6G 2E9
(403) 492-3308
FAX (403) 492-9234
bio.grad.coordinator@ualberta.ca

Department of Biological Sciences
Simon Fraser University
Burnaby, BC
Canada V5A 1S6
(604) 291-4475
FAX (604) 291-3496
http://biol.sfu.ca

Department of Entomology
University of Manitoba
Winnipeg, MB
Canada R3T 2N2
(204) 474-9257
FAX (204) 474-7628
neil holliday@unamitoba.ca

Department of Biology
Memorial University
St. Johns, NF
Canada A1B 3X9
(709) 737-7497
FAX (709) 737-3018

Department of Zoology
University of Western Ontario
London, ON
Canada N6A 5B7

Collections & Research
Canadian Museum of Nature
P. O. Box 3443, Station D
Ottawa, ON
Canada K1P 6P4
(613) 364-4060
FAX (613) 364-4027
randerson@mus-nature.ca
fgenier@mus-nature.ca

Eastern Cereal and Oilseed Research Centre
Biological Resources Program
Agriculture and Agri-Food Canada
Ottawa, ON
Canada K1A 0C6
(613) 759-1864
FAX (613) 759-1924
mckenziejs@em.agr.ca

Entomology Section
Centre for Biodiversity and Conservation Biology
Royal Ontario Museum
100 Queen's Park
Toronto, ON
Canada M5S 2C6
(416) 586-8059
FAX (416) 586-5553
chrisd@rom.on.ca

Department of Zoology
University of Toronto
Toronto, ON
Canada M5S 3G5
(416) 978-2084
FAX (416) 978-8532
hala@zoo.utoronto.ca

Department of Natural Resource Sciences
Faculty of Agricultural and Environmental Sciences
McGill University (Macdonald Campus)
21,111 Lakeshore Road
Sainte-Anne de Bellevue, Quebec
Canada H9X 3V9
(514) 398-7890
FAX (514) 398-7990
teneyck@agradm.lan.mcgill.ca

Lyman Entomological Museum & Research Laboratory
Macdonald Campus of McGill University
St. Anne de Bellevue, Quebec
Canada H9X 1C0
(514) 398-7914
FAX (514) 398-7990
lyman@nrs.mcgill.ca

Department of Biology
University of Saskatchewan
112 Science Place
Saskatoon, SK
Canada S7N 5E2
(306) 966-4400
FAX (306) 966-4461
gillott@duke.usask.ca

Esp. de Entomologia y Acarologia
Instutito de Fitosanidad
Colegio de Postgraduados
C.P. 56230
Montecillo, Texcoco, Edo. de Mexico
Mexico
(595) 1-15-80 & 1-02-20
FAX (595) 1-15-80 & 1-02-20

Graduados en Agricultura
ITESM, Suc. Correos "J"
Monterrey, N.L. 64849
Mexico
(8) 358-20-00 ext. 5190
FAX (8) 359-92-06
earanda@campus.mty.itesm.mx

Colegio De Graduados De La Facultad De Agron
Universidad Autonoma De NuevoLeon
APDO.Postal No. 358
SN. Nicolas De Los Garza
Nuevo Leon
Mexico
(824) 8-01-01 y 8-02-59
FAX (824) 8-01-01
fzavala@crr.dsi.uanl.mx

Index

C

D

H

I

J

K

L

T